Rayner Joel
B.Sc.(Eng.) Lond., C.Eng., F.I.M

Thermodynamics
Level 4

Longman London and New York

Longman Group Limited
Longman House, Burnt Mill, Harlow
Essex CM20 2JE, England
Associated companies throughout the world

*Published in the United States of America
by Longman Inc., New York*

© Longman Group Limited 1985

All rights reserved; no part of this publication may be
reproduced, stored in a retrieval system, or transmitted
in any form or by any means, electronic, mechanical,
photocopying, recording, or otherwise, without the
prior written permission of the Publishers.

First published 1985

British Library Cataloguing in Publication Data

Joel, Rayner
  Thermodynamics level 4.
    1. Thermodynamics   2. Heat engineering
  I. Title
  621.402′1   TJ265

ISBN 0582-41300-1

Set in Monophoto Times
Produced by Longman Singapore Publishers (Pte) Ltd.
Printed in Singapore

General Editors – Mechanical and Production Engineering

H. G. Davies
Vice Principal and Head of Department of Science, Carmarthen Technical and Agricultural College

G. A. Hicks
Lecturer in the Department of Engineering, Carmarthen Technical and Agricultural College

Books published in this sector of the series:

Workshop processes and materials Level 1 Second edition
    R. L. Timings
Manufacturing technology Level 2 Second edition   *R. L. Timings*
Manufacturing technology Level 3 Second edition   *R. L. Timings*
Engineering science for mechanical technicians Level 2   *J. O. Bird and A. J. C. May*
Engineering science for mechanical technicians Level 3   *J. O. Bird and A. J. C. May*
Engineering drawing for technician engineers   *J. D. Poole*
Engineering drawing for technician engineers Level 2   *J. D. Poole*
Thermodynamics Level 3   *Rayner Joel*
Materials technology Level 2   *R. L. Timings*
Materials technology Level 3   *R. L. Timings*

Contents

Preface

Chapter 1	The First Law of Thermodynamics	1
Chapter 2	The Second Law of Thermodynamics	4
Chapter 3	Thermodynamic reversibility	8
Chapter 4	Entropy	17
Chapter 5	Gas and vapour power cycles	65
Chapter 6	Refrigerators and heat pumps	141

Index 158

Preface

This book has been primarily written to cover the syllabus aims of the Business and Technician Education Council (BTEC) Level 4 unit of thermodynamics. This unit follows on from the Level 3 unit of thermodynamics.

The first and second laws of thermodynamics are explored and the concepts of thermodynamic reversibility and of entropy are introduced. They will be seen as useful ideas which have led to the further understanding and practical application of thermodynamics. The translation into practice is introduced in the chapters on gas and vapour power cycles and also refrigerators and heat pumps.

The book has been arranged in chapters which should preferably be read in sequence.

R.J
Leigh-on-Sea 1984

Chapter 1

The First Law of Thermodynamics

The First Law of Thermodynamics is covered in detail in the book *Thermodynamics Level 3*.

A summary of the First Law is now given.

1.1 The First Law of heat and work transfer

Work transfer and heat transfer are related by the First Law.

The same effect can be produced by work transfer as can be produced by heat transfer on a system.

Starting at a given temperature, both transfers can be made to produce the same temperature change and thus, in energy terms,

Work transfer = Heat transfer

or

$$W = Q \tag{1}$$

Equation (1) is a statement of the First Law of Thermodynamics.

1.2 The First Law and the cycle

If a substance passes through a series of processes such that it is eventually returned to its original state then the substance is said to have been taken through a cycle. Since, after a cycle, a substance is returned to its original state, then the algebraic sum of the energy transfers which took place during the cycle must be zero.

For this to be so, then, for a cycle

Net work − Net heat = 0   (2)

or

Net work = Net heat   (3)

or

$$\sum W = \sum Q \quad (4)$$

or

$$\oint W = \oint Q \quad (5)$$

Equations (2), (3), (4) and (5) are further statements of the First Law of Thermodynamics.

Note that the First Law implies that, for a cycle, there must be heat transfer for there to be work transfer.

For example, a machine which could provide work transfer without heat transfer would violate the First Law because it would create energy. This is contrary to the principle of conservation of energy.

No violation of the First Law has been shown.

Of interest, a machine which could provide work transfer without heat transfer would run forever, or in other words, it would have perpetual motion!

Such a machine would have what is sometimes referred to as perpetual motion of the first kind.

Another impossible perpetual motion machine is discussed in the Second Law of Thermodynamics (Ch. 2).

1.3 The first law and the non-flow energy equation

If a system does not execute a cycle then the algebraic sum of work and heat energy transfers does not equal zero. Thus

$$\sum W \neq \sum Q \quad (6)$$

By the principle of conservation of energy

Energy in = Energy out   (7)

If the work and heat transfers are not equal then the energy difference must have been added to the substance or have been a loss from the substance.

This introduces the concept of *internal energy*.

The energy residing in the substance is called the internal energy, U.

With the inclusion of internal energy, equation (1) becomes

$$Q = \Delta U + W \quad (8)$$

where

Q = heat transfer

ΔU = change in internal energy

W = work transfer

Equation (8) neglects local internal gravitational and kinetic energies.

This equation (8) is the *non-flow energy equation* and is yet another statement of the First Law of Thermodynamics.

1.4 The first law and the steady-flow energy equation

For a steady flow process, the First Law develops into the *steady flow energy equation*, which for specific quantity becomes

$$gZ_1 + u_1 + P_1v_1 + \frac{C_1^2}{2} + Q = gZ_2 + u_2 + P_2v_2 + \frac{C_2^2}{2} + W \qquad (9)$$

where

Z = height above given datum (gives gravitational potential energy)

u = internal energy

P_1v_1 = flow work

C = velocity (gives kinetic energy)

Q = heat transfer

W = work transfer

Note the equations (8) and (9) are forms of *energy balance*.

Chapter 2

The Second Law of Thermodynamics

2.1 The Second Law and Temperature Gradient

The Second Law of Thermodynamics is a directional law and states that heat transfer will occur of its own accord down a temperature gradient as a natural phenomenon.

Heat trasnfer can be made to transfer up a temperature gradient but not without the aid of external energy.

Natural heat transfer down a temperature gradient degrades energy to a less valuable level. A limit of value occurs when temperatures become equal and are thus in equilibrium.

2.2 The second law and heat and work transfer

From the First Law of Thermodynamics, for a cycle, and hence engines because engines must work in cycles to continue operating,

Net work transfer = Net heat transfer (1)

or $\sum W = \sum Q$ (2)

Now experience always shows, however, that

Net work transfer < Net heat transfer (3)

or $\sum W < \sum Q$ (4)

Since the work transfer is less than the heat transfer,

$$\sum Q - \sum W > 0 \tag{5}$$

and has some positive value.

This means that some heat transfer must be rejected and is lost. Therefore there must always be some inefficiency.

Further, there is no work transfer unless there is a temperature difference because as shown in section 2.1 unless there is a temperature difference there is no heat transfer and equations (1), (2), (3), (4) and (5) require that there is heat transfer in order that there shall be some work transfer.

2.3 Statements of the second law

The concepts of the Second Law of Thermodynamics enunciated in sections 2.1 and 2.2 were put together in the past in other forms, notably as follows:

Sadi Carnot (1796–1832)
'Whenever a temperature difference exists, motive power can be produced.'

Strictly, this was not given as a statement of the Second Law of Thermodynamics. Carnot's concept of heat was in error (see *Thermodynamics*, Level 3).

Carnot's concept of heat was in error in that, at that time, heat was thought to have the properties of a fluid which flowed into or out of a body. It was suggested that as the result of this 'fluid' flow, a body became hotter or colder. Heat was not considered an energy transfer process occurring naturally as the result of temperature difference which is now the accepted way by which heat is defined. (See *Thermodynamics* Level 3, sections 1.16 and 1.19.)

However, the suggestion that temperature difference is the prerequisite of the ability to produce motive power is correct.

It is also a positive statement in that it declares when it is possible to produce motive power.

The statements which follow are negative because they declare impossibilities.

Rudolf Clausius (1822–1888)
'It is impossible for a self-acting machine, unaided by any external agency, to convey heat from a body at a low temperature to one at a higher temperature.'

Note that the implication here is that unless external energy is made available, heat transfer is impossible up a gradient of temperature.

The fact that heat transfer can be made to occur up a temperature gradient is made manifest by the refrigerator. However, the refrigerator is not self-acting. It requires external energy in order that it can operate.

William Thompson, later Lord Kelvin (1824–1907)

'We cannot transform heat into work merely by cooling a body already below the temperature of the coldest surrounding objects.'

This implies that when a body reaches the temperature of the coldest surrounding objects no further natural heat transfer is possible and hence no further work transfer is possible.

Max Planck (1858–1947)

'It is impossible to construct a system which will operate in a cycle, extract heat from a reservoir, and do an equivalent amount of work on the surroundings.'

This statement implies the inability to completely convert heat transfer into work transfer. The inference is that there must always be some heat transfer rejection which is a loss from the system.

Kelvin–Planck

'It is impossible for a heat engine to produce net work in a complete cycle if it exchanges heat only with bodies at a single fixed temperature.'

This combination of individual statements implies that it is not possible to produce work transfer if a system is connected only to a single heat energy source or reservoir which is at a single fixed temperature.

2.4 Implications of the second law

No contradiction of the Second Law of Thermodynamics has been demonstrated.

The main implications of the Second Law of Thermodynamics are:

1. Heat transfer will only occur, and will always naturally occur, when a temperature difference exists and always naturally down the temperature gradient.
2. If there is heat transfer availability, due to temperature difference, work transfer is always possible. However, there is always some heat transfer loss.
3. Temperature can be elevated but not without the expenditure of external energy. Elevation of temperature cannot occur unaided.
4. There is no possibility of work transfer if only a single heat energy source or reservoir at a fixed temperature is available.
5. It should be noted that if work transfer is supplied to a system then

this can all be transformed into heat energy. Examples of this are friction and the generation of electrical energy.

Heat energy, however, cannot all be transformed into work transfer. There will always be some loss.

Thus work transfer has a higher energy transfer value than heat transfer.

It is important to attenuate this last statement (5) however, because usually work transfer is only made available by the expenditure of heat transfer.

From this Second Law of Thermodynamics then, it follows that in order to run all the engines and devices in use today, and hence maintain and develop modern industrial society, a source of supply of suitable fuels is absolutely essential. It is by burning and consuming fuel that the various working substances have their temperature put up above that of their surroundings thus enabling them to release energy by heat transfer in a natural manner according to the Second Law.

By virtue of the Second Law of Thermodynamics it is essential that all fuels should be used as efficiently as possible in order that fuel stocks may be preserved for as long as possible. It must always be remembered that when once energy has been degraded by heat transfer down a temperature gradient then further energy is only made available at the expense of further fuel.

Chapter 3

Thermodynamic reversibility

3.1

If a substance passes through a process in such a manner that, after the process, the substance can be taken back through all the stages which are passed through in a reversed order until it finally reaches its original state, then the process is said to be reversible.

After carrying out a reversible process, there would be no evidence anywhere that the process had ever taken place.

No such process exists in practice.

Within the substance during any process it is probable that eddies will be set up. Also, due to the viscosity of the substance, however slight, there will be some internal friction. It is also very likely that there will be some small irregularity with regard to the distribution of temperature throughout the substance. The degree to which these occur must have some bearing on the final state of the substance after the process. From here, however, it is unreasonable to assume that these various internal phenomena can be repeated in an exactly reversed sequence in order that the reversed process will return the substance to its original state. For these reasons alone no actual process can be considered as truly reversible.

A point to raise here, however, is that the effect of these internal phenomena is not likely to be great and from a theoretical standpoint it is possible to neglect them.

This being so then, it is necessary to move on to consider whether there is anything else which will affect the concept of reversibility.

As an example, consider the expansion of a gas.

During an expansion of a gas, with the exception of the adiabatic case in which, by definition, there is no heat transfer (see *Thermodynamics* Level 3, section 1.33), there will be heat transfer into or out of the gas in all cases.

Now the Second Law of Thermodynamics states that heat transfer will only occur down a temperature gradient as a natural occurrence.

During the expansion, assume that there is some heat transfer from the gas to the surroundings.

If this is the case, then, by the Second Law of Thermodynamics, the surroundings are at a lower temperature than the gas.

What if an attempt is made to reverse the process?

This now means compressing the gas, which is easy enough. However, it is not possible to return the energy lost by heat transfer to the surroundings because the gas is at a higher temperature than the surroundings.

Certainly it would be possible to return the gas to its original volume.

The original pressure and temperature could not be attained, however, because of the energy lost during the original expansion by heat transfer to the surroundings. This energy loss cannot be returned because of the limitations of the Second Law of Thermodynamics.

Notice also that there would also be some heat transfer loss during the reversed process of compression because again the temperature of the gas would be above that of the surroundings.

A similar analogy would hold good in the case of an expansion in which heat transfer occurred into the gas. Once again, by the Second Law of Thermodynamics, the reversed heat transfer is impossible because the surroundings would be at a higher temperature than the gas in order that the original heat transfer could occur.

Another point to consider is the effect of pressure imbalance. If a substance at a high pressure expands into surroundings which are at a lower pressure then the reversed process of the low-pressure surroundings returning the substance to its original high pressure is impossible without the aid of external energy. Thus if pressure imbalance occurs then a process cannot be reversible.

From this then, it follows that the great majority of thermodynamic processes are irreversible.

Now irreversibility evidently involves loss, and hence it appears that reversibility is bound up with efficiency. A truly reversible process involves no loss and hence is the most efficient thermodynamic process possible.

No external energy is required to return a substance to its original state in a truly reversible process.

It is important, therefore, to investigate whether there are any processes which may be considered as being theoretically reversible.

3.2 The adiabatic process

Here, no heat is transferred during the process. Thus the effect of the Second Law of Thermodynamics between the substance and its surroundings is eliminated.

If the effects of pressure imbalance, internal friction, non-uniform temperature distribution, etc., are neglected then it follows that the adiabatic process is theoretically reversible.

For an adiabatic process it is shown that $W = -\Delta U$. Thus, during an adiabatic expansion external work is done which equals the decrease in internal energy. If, now, this same amount of work is done on the working substance during the reversed process of adiabatic compression then the work will appear as an increase of internal energy of the substance and this increase will just equal the loss which occurred during the expansion. Thus the substance will be returned exactly to its original state.

The adiabatic process is, therefore, theoretically reversible.

3.3 The isothermal process

This process is carried out at constant temperature.

For an isothermal, non-flow process, it is shown that the necessary energy exchange is that $Q = W$.

This means that, during an isothermal expansion, the working substance must receive an amount of heat equal to the external work done. It follows that if, during the reversed process of compression, an amount of heat equal to the work done on the substance is rejected then, neglecting the effects of pressure imbalance, internal friction, non-uniform temperature distribution, etc., the isothermal process is theoretically reversible.

With regard to the transfer of heat into or out of the substance it must be remembered that an isothermal process is carried out at constant temperature. Assuming that the external surroundings are at this temperature then heat transfer is equally possible in either direction, namely, into or out of the substance. Actually, by the Second Law of Thermodynamics, a temperature difference is required in order to promote a natural heat transfer. Therefore, during an isothermal expansion, it could be considered that the surroundings are at a slightly elevated temperature above the substance and hence the necessary condition that the substance shall receive heat would be met. Similarly, it could be considered that the substance has a slightly elevated temperature above the surroundings during an isothermal compression. In this case the necessary condition that the substance should reject heat would be met. Since the temperature difference in each case would be small then the rate of heat transfer would be very slow, and hence, from a practical point of view, the isothermal process is very slow.

Actually it is all but impossible as a practical process. However, theoretically it exists and, neglecting the effect of internal friction, etc., it is theoretically reversible.

3.4 The polytropic, constant volume and constant pressure processes

In all these cases heat is received or rejected by the substance during the progress of the process and also the temperature changes continually throughout the process. If the temperature of the surroundings remains constant then, by the Second Law of Thermodynamics, heat transfer between the substance and surroundings is uni-directional, being a function of whether the substance is at a higher or lower temperature than the surroundings.

Also, apart from the constant pressure process in which the pressure of the substance could be made the same as the pressure of the surroundings, in these processes where pressure interaction occurs between the substance and the surroundings, there is pressure imbalance.

In these cases, then, in which there is pressure and temperature difference between the substance and surroundings, the processes are irreversible.

These processes could be considered as reversible if the temperature and pressure of the surroundings could be made to vary in the same way as the temperature and pressure of the substance. In this way, in a similar manner to the isothermal case, mutual heat transfer in either direction would be possible, and there would be no pressure imbalance. The processes could then be considered as being reversible. In the main, these conditions are impossible to achieve in practice.

3.5 The non-flow energy equation and reversibility

The non-flow energy equation connecting the initial and final states of a substance has the form,

$$Q = (U_2 - U_1) + W$$

This equation can be used for either a reversible or irreversible process so long as the initial and final states are in equilibrium, i.e., there is pressure and temperature equilibrium throughout the substance and no random internal energies exist due to such things as turbulence. Also an exact knowledge of the true amount of external work done must be known.

If, however, the differential form of this equation is used,

$$dQ = dU + dW$$

then this only applies to a reversible process. To explain this, firstly consider the work done.

For calculation, the work done is determined by integrating the expression,

$$dW = P\,dV$$

To use this expression it must be assumed that the pressure P is always resisted by an opposing pressure equal to P. To clarify this, consider a cylinder-piston arrangement. If the substance pressure is higher than the opposing pressure, produced by an external load upon which work is being done, then there will be less work done than $\int P\,dV$ would indicate. In the extreme case when the expansion is free, such as when a substance expands into a vacuum, as in Joule's Internal Energy Experiment (see *Thermodynamics Level 3*, section 4.4), no external work is done even though a change in pressure and volume occurs. In any event, considering the reverse process in which external work is required to compress the substance, a lower external pressure cannot be made to compress a substance already at a higher pressure. If, however, the pressure, both internally and externally is the same at all times, then the compression is theoretically refersible and $\int P\,dV$ gives the external work done exactly.

With regard to the change of internal energy given by $\int dU$ this is only valid if the substance has passed through a series of equilibrium stafes and thus there have been no random internal energies present, such as turbulence. This is only possible in a reversible case.

Now since the heat transfer during a non-flow process is dependent upon the change of internal energy and the external work done and since, to use the expression,

$$dQ = dU + dW$$

dU and dW must be reversible, then it follows that dQ must also be reversible. For this to be so it implies that the external surrounding temperature must vary exactly with the substance temperature, as already explained earlier.

From the above discussion, since the heat transfer during a non-flow process is usually calculated using the expression,

$$Q = \int dU + \int dW$$

$$= \int dU + \int P\,dV$$

then this calculated heat transfer is for a reversible case only.

To indicate this, the heat transfer is written Q_{rev}.

Thus,

$$Q_{rev} = \int dU + \int P\,dV$$

If, for reversible heat transfer, the external temperature must vary exactly with the substance temperature, then, as explained in the isothermal case, the process would be very slow. In fact, to operate at all, the process must be theoretically an infinitely slow process. All practical processes take a finite time and thus, from this point of view alone, become irreversible.

From the discussion, it follows that calculations of heat transfer, change of internal energy and external work done are close approximations only, the calculated results applying only to ideal reversible cases.

3.6 Carnot's Principle

It has already been stated that a reversible thermodynamic process is the most efficient thermodynamic process because it involves no loss of energy.

Reversibility, as it applies to the thermodynamic engine, was discussed by a Frenchman, Sadi Carnot, in a paper entitled 'Reflections on the motive power of heat' which was published in 1824. In the paper, Carnot conceived an engine working on thermodynamically reversible processes, and from this concept deduced what has since been called *Carnot's Principle*. This states that 'no engine can be more efficient than a reversible engine working between the same limits of temperature'.

The principle of the thermodynamic engine is that it receives heat at some high temperature from a heat source. The engine then converts some of this heat into work and then rejects the remainder into a sink.

Consider, then, a thermodynamically reversible engine R working between the temperature limits of source T_1 and sink T_2.

In some period of time let this engine receive Q units of heat from the source at temperature T_1. It will convert W_R units of this heat into work and then reject $(Q - W_R)$ units of heat into the sink at lower temperature T_2. This is shown in Fig. 3.1(a).

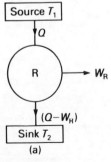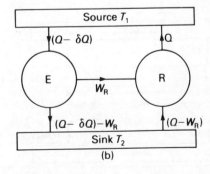

Fig. 3.1

Now assume that some other engine E can be found which is more efficient than the reversible engine R. Since it is more efficient, this engine E will require less heat supplied to perform the same amount of work W_R. Let this engine E drive engine R reversed and let them both work between the same source and sink. This is shown in Fig. 3.1(b).

Engine E, being more efficient, will require $(Q - \delta Q)$ units of heat supplied from the source at temperature T_1. It will convert W_R of this into work and thus it will reject $(Q - \delta Q) - W_R$ units of heat into the sink at temperature T_2.

Now the work W will drive engine R reversed, which now becomes a heat pump. Thus it will take up $(Q - W_R)$ units of heat from the sink at lower temperature T_2. It will convert W_R units of work into heat and then reject $(Q - W_R) + W_R = Q$ units of heat into the source at higher temperature T_1.

Investigation of this system will show that, during the time period considered, there has been a gain of heat to the source $= Q - (Q - \delta Q) = \delta Q$ units of heat.

Also the sink has lost $(Q - W_R) - \{(Q - \delta Q) - W_R\} = \delta Q$ units of heat.

This means that the source at higher temperature T_1 is receiving heat from the sink at lower temperature T_2.

Now this arrangement is self-acting and has apparently managed to make more heat transfer up the gradient of temperature than has moved down. This would mean that eventually all the heat would be transferred to the source at temperature T_1 while the sink at lower temperature T_2 would have its energy content reduced to zero!

This is contrary to the Second Law of Thermodynamics and thus the system is impossible.

If, however, the engine E has the same efficiency as the reversible engine R, then Fig. 3.2 shows that the thermodynamic system balances in which case both the source and sink gain as much heat as they lose.

This means that the energy level of both the source and the sink would remain constant. This system would thus, once started, continue to run indefinitely and hence it would have perpetual motion!

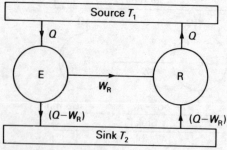

Fig. 3.2

No such system exists, since no engine can be made to have the same efficiency as that of a reversible engine. In any case, the above system has assumed no loss which, again, is impossible. The important criterion, however, which has been established is that the thermodynamically reversible engine has the maximum efficiency possible.

An alternative analysis showing that the thermodynamically reversible engine has the maximum efficiency possible is as follows.

Again, consider a reversible engine R and assume that there exists an engine E which has a higher efficiency than reversible engine R.

Consider Fig. 3.3(a). In this case let engine E receive heat Q from the source at temperature T_1. It will produce work W_E which is greater than W_R because it is more efficient. It will reject heat $(Q - W_E)$ to the sink at temperature T_2.

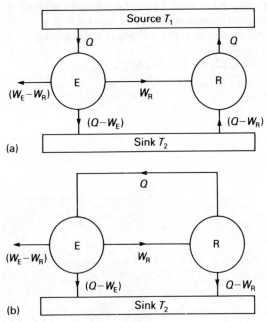

Fig. 3.3

Engine E is arranged to drive engine R reversed.

Engine R will require work W_R ($W_R < W_E$) to drive it and will take up heat $(Q - W_R)$ from the sink at temperature T_2 while it rejects heat Q into the source at temperature T_1.

Now since the reversible engine R rejects heat Q which is exactly the requirement of engine E, then, theoretically it would be possible to dispense with the heat source at temperature T_1. This is shown in Fig. 3.3(b).

Consider this new arrangement. Here, apparently, is a self-acting machine which is producing a net work output $=(W_E - W_R)$, while exchanging heat with only a single reservoir, in this case the sink at temperature T_2.

This is contrary to the Second Law of Thermodynamics as stated in the combined concept of Kelvin–Planck, section 2.3.

This analysis once again shows that the thermodynamically reversible engine has the maximum efficiency possible within the given limits of temperature.

For the given temperature limits of source T_1 and sink T_2 it is now necessary to determine this maximum efficiency. To establish it, all processes must be thermodynamically reversible and the arrangement of the processes must be such that they are capable of cyclic repetition in order that an engine, when started, may continue to run by the process of repetition of the *cycle*, as it is called.

Carnot conceived a cycle made up of thermodynamically reversible processes. By determining the thermal efficiency of this cycle it is possible to establish the maximum possible efficiency between the temperature limits of the cycle.

The Carnot cycle will be analysed in Chapter 5.

Chapter 4

Entropy

4.1 Introduction

During many non-flow processes it is necessary to investigate the heat transferred during the progress of the process.

The heat transferred during a process will affect any work transfer which may occur during the process and also the end state after the process.

Further, it has been shown that the theoretical amount of heat transferred, determined by calculation, is transferred reversibly. From here it might be suggested that if a graph could be developed such that the area underneath a process plotted on the graph gave the amount of heat transferred reversibly during the progress of the process, then such a graph would perform a useful function. The idea is analogous to the area of the pressure-volume graph which gives the work done during a reversible process.

The problem now is to decide what the axes of the graph are to be.

Let one axis be absolute temperature T and the other some new function s as shown in Fig. 4.1(a).

Absolute temperature is chosen as one axis since it has a very close relationship with the energy level of a substance, notably internal energy and enthalpy.

Also, when a substance is at the absolute zero of temperature, it is assumed that its internal energy content is also zero.

It will also be remembered that heat is defined as that energy transfer which will occur as the result of a temperature difference.

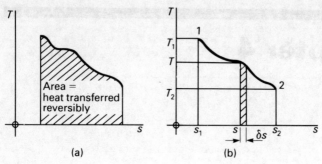

Fig. 4.1

Now consider Fig. 4.1(b) in which is shown a process plotted on the graph where a change has occurred from state 1 to state 2. Consider some point on this graph where the coordinates are T and s. Let the state change from this point such that there is a small change in $s = \delta s$, then,

Heat transferred reversibly = Area swept out by this small change

$$= T\,\delta s \text{ (very nearly)}$$

From this,

Heat transferred reversibly from 1 to 2

= Total area under graph from 1 to 2

$$= \sum_{S=S_1}^{S=S_2} T\,\delta s$$

In the limit as $\delta s \to 0$,

Heat transferred reversibility from 1 to 2

$$= \int_{S_1}^{S_2} T\,\mathrm{d}s = Q_{\text{rev}} \tag{1}$$

Differentiating equation (1)

$$\mathrm{d}Q_{\text{rev}} = T\,\mathrm{d}s$$

or

$$\mathrm{d}s = \frac{\mathrm{d}Q_{\text{rev}}}{T} \tag{2}$$

This equation gives the relationship which must exist between s, T and Q_{rev} in order that the area of the graph shall be heat transferred reversibly.

Now it has already been shown that it is possible to calculate the amount of heat transferred reversibly during a non-flow process. Thus, by using equation (2) changes in s can be determined. It is this function s which is called *entropy*.

Inspection of equation (2) will show that if heat is received, which makes Q_{rev} positive, then the entropy of the receiving substance has increased.

Conversely, if heat is rejected, which makes Q_{rev} negative, then the entropy of the rejecting substance is decreased.

Thus, positive and negative changes of entropy show whether heat has been received or rejected during the process considered.

Now consider an isolated system in which an amount of heat energy Q is transferred from a hot source at temperature T_1 into a cooler sink at temperature T_2.

The loss of entropy from the hot source is Q/T_1 while the gain of entropy to the cooler sink is Q/T_2.

The amount of heat transferred is the same for both the source and the sink but, since $T_1 > T_2$, it follows that the gain of entropy to the cooler sink is greater than the loss of entropy from the hot source, since,

$$\frac{Q}{T_2} > \frac{Q}{T_1} \tag{3}$$

Now, by the Second Law of Thermodynamics, heat transfer will only occur down a temperature gradient as a natural occurrence.

This makes the natural transfer of heat an irreversible process.

Thus it appears, that if a process occurs in an isolated system such that there is an increase in entropy then the process is irreversible.

This leads to a statement of the Principle of Increase of Entropy. This is that an isolated system can only change to states of equal or greater entropy, or,

$$\Delta s \geqslant 0 \tag{4}$$

where $\Delta s =$ change of entropy.

Note, also, that the higher the temperature of a system is above that of the surroundings, then the greater becomes the availability of the energy obtained by heat transfer.

Since entropy is a function of temperature (see equation 2), then it follows that entropy is associated with the usefulness of energy.

The method of calculation of changes of entropy will now be investigated. The expressions for change of entropy are developed assuming the processes to be reversible. It must be remembered, however, that a change is dependent only upon the end states and not on how the change occurred. Thus the expressions for change of entropy can also be used for irreversible processes.

Note, from equation (2), if unit mass of substance is considered, the unit of specific entropy becomes J/kg K or, very often, kJ/kg K.

4.2 The entropy of vapours (two-phase systems)

As with enthalpy, so entropy is treated separately in the three stages of the formation of a vapour from a liquid.

4.3 Liquid entropy

Consider unit mass of a liquid which will ultimately be changed into unit mass of vapour at constant pressure. For the unit mass of liquid,

$$dQ = c_{pL}\, dT \tag{1}$$

where c_{pL} = specific heat capacity of the liquid at constant pressure.

Dividing equation (1) throughout by T, then,

$$\frac{dQ}{T} = c_{pL}\frac{dT}{T} = ds \tag{2}$$

since $ds = dQ/T$.

∴ For this case,

$$ds = c_{pL}\frac{dT}{T} \tag{3}$$

Integrating this equation from initial state 1 to final state 2

$$\int_{s_1}^{s_2} ds = c_{pL}\int_{T_1}^{T_2}\frac{dT}{T}$$

or

$$\left[s\right]_{s_1}^{s_2} + c_{pL}\left[\ln T\right]_{T_1}^{T_2}$$

from which,

$$s_2 - s_1 = c_{pL}[\ln T_2 - \ln T_1]$$

or

$$s_2 - s_1 = c_{pL}\ln\frac{T_2}{T_1} \tag{4}$$

In the same way as the zero of internal energy and enthalpy are arbitrarily chosen, so the zero of entropy is arbitrarily chosen.

In the case of steam, this arbitrary zero is chosen at the triple point, whose temperature is 273.16 K (273 K is accurate enough for most calculations).

Hence, in equation (4) let

$s_1 = 0$ when $T_1 = 273.16$ K

then,

$$s_2 - 0 = c_{pL} \ln \frac{T_2}{273.16}$$

or, dropping the suffix 2,

$$s = c_{pL} \ln \frac{T}{273.16} \tag{5}$$

If it is assumed, that, for water,

$c_{pL} = 4.187$ kJ/kg K

then equation (5) becomes,

$$s = 4.187 \ln \frac{T}{273.16} \tag{6}$$

Actually, c_{pL} for water varies with temperature, but 4.187 kJ/kg K is an average value at normal low range temperatures.

When saturation temperature is reached then maximum liquid entropy is reached for the particular pressure under consideration. The specific entropy is then writteh s_f.

Thus, at saturation temperature for water,

$$s_f = c_{pL} \ln \frac{T_f}{273.16} \tag{7}$$

or, if c_{pL} assumed $= 4.187$ kJ/kg K

$$s_f = 4.187 \ln \frac{T_f}{273.16} \tag{8}$$

Units will be kJ/kg K.

Example 1

Determine the value of the specific entropy of water at 143.6 °C. From equation (8):

$$s_f = 4.187 \ln \frac{T_f}{273.16} = 4.187 \ln \frac{416.6}{273.16}$$

$$= 4.187 \ln 1.525 = 4.187 \times 0.422$$

$$= \underline{1.767 \text{ kJ/kg K}}$$

From tables, the accurate value of s_f in this case $= \underline{1.776 \text{ kJ/kg K}}$

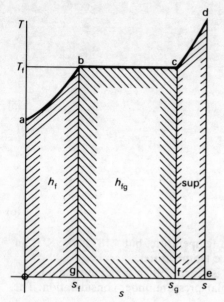

Fig. 4.2

4.4 Entropy of evaporation

Consider Fig. 4.2. Here is shown a temperature-entropy (T-s) diagram of the formation of vapour at constant pressure. Curve ab represents the introduction of the liquid enthalpy to the water. At b the water reaches saturation temperature t_f. Horizontal line bc represents the introduction of the enthalpy of evaporation at constant temperature t_f. There is all liquid at b and all dry saturated vapour at c. Curve cd represents the introduction of the superheat. Note that line abcd has a similar appearance to that obtained in the temperature-enthalpy diagram.

Now it has been shown that the area under a T-s diagram gives heat transferred (reversibly). Also, at constant pressure,

Heat transferred = Change of enthalpy

Hence from Fig. 4.2,

Area abgo = h_f = Specific liquid enthalpy
Area bcfg = h_{fg} = Specific enthalpy of evaporation
Area cdef = sup = Specific superheat

From the diagram

Area bcfg = $h_{fg} = T_f(s_g - s_f)$

where s_g = Specific entropy of dry saturated steam.

Hence

$$s_g - s_f = \frac{h_{fg}}{T_f} = \text{Specific entropy of evaporation} = s_{fg} \tag{9}$$

Also, from equation (9)

$$s_g = s_f + s_{fg} \tag{10}$$

$$= s_f + (s_g - s_f) \tag{11}$$

From equation (10), in the case of steam,

$$s_g = c_{pL} \ln \frac{T_f}{273.16} + \frac{h_{fg}}{T_f} \tag{10}$$

Accurate values of s_g and s_{fg} are given in tables.
If the vapour formed is wet, then the specific enthalpy of evaporation introduced if the dryness fraction is $x = xh_{fg}$.
If the specific entropy of the wet vapour $= s$, then,

$$s = s_f + x \frac{h_{fg}}{T_f} = s_f + x s_{fg} \tag{13}$$

$$= s_f + x(s_g - s_f) \tag{14}$$

Example 2
Determine the value of the specific entropy of wet steam at a pressure of 1.5 MN/m² (1.5 MPa) and 0.9 dry:

(a) by calculation;
(b) by using values of entropy from steam tables.

(Note: at 1.5 MN/m², $t_f = 198.3\ °C$.
∴ $T_f = 198.3 + 273 = 471.3\ K$.)

(a) $s = c_{pL} \ln \dfrac{T_f}{273.16} + x \dfrac{h_{fg}}{T_f}$

$= 4.187 \ln \dfrac{471.3}{273.16} + 0.9 \times \dfrac{1\,945.2}{471.3}$

$= 4.187 \times \ln 1.725 + 3.715$

$h = 4.187 \times 0.545 + 3.715$

$= 2.282 + 3.715$

$= \underline{5.997\ kJ/kg\ K}$

(b) $s = s_f + x s_{fg}$

$= 2.314 + 0.9 \times 4.127$

$= 2.314 + 3.714$

$= \underline{6.028 \text{ kJ/kg K}}$

This is the accurate value.

4.5 Entropy of superheated vapour

Let c_{pv} = specific heat capacity of superheated vapour at constant pressure. Heat received in the superheat region

$= dQ = c_{pv} dT$

Hence, for the superheated vapour,

$$\frac{dQ}{T} = ds = c_{pv} \frac{dT}{T}$$

Integrating this equation from saturation temperature T_f to superheated steam temperature T, then,

$$\int_{s_g}^{s} ds = c_{pv} \int_{T_f}^{T} \frac{dT}{T}$$

or

$$\left[s \right]_{s_g}^{s} = c_{pv} \left[\ln T \right]_{T_f}^{T}$$

hence

$s - s_g = c_{pv} [\ln T - \ln T_f]$

$\therefore \quad s - s_g = c_{pv} \ln \dfrac{T}{T_f}$

from which

$$s = s_g + c_{pv} \ln \frac{T}{T_f} \tag{15}$$

or, for superheated steam,

$$s = c_{pL} \ln \frac{T_f}{273.16} + \frac{h_{fg}}{T_f} + c_{pv} \ln \frac{T}{T_f} \tag{16}$$

Accurate values of s for superheated vapours are given in tables.

Example 3
Determine the value of the specific entropy of steam at 1 MN/m² (1.0 MPa) and with a temperature of 275 °C:
(a) by calculation;
(b) from steam tables.

Take $c_{pv} = 2.3$ kJ/kg K.

(a) $s = c_{pL} \ln \dfrac{T_f}{273.16} + \dfrac{h_{fg}}{T_f} + c_{pv} \ln \dfrac{T}{T_f}$

$= 4.187 \ln \dfrac{452.9}{273.16} + \dfrac{2013.6}{452.9} + 2.3 \ln \dfrac{548}{452.9}$

$= 4.187 \ln 1.658 + 4.446 + 2.3 \ln 1.21$

$= (4.187 \times 0.506) + 4.446 + (2.3 \times 0.191)$

$= 2.119 + 4.446 + 0.439$

$= \underline{7.004 \text{ kJ/kg K}}$

(b) From tables, $s = \underline{7.029 \text{ kJ/kg K}}$

This is the accurate value.

4.6 The temperature-entropy chart for vapours

Extending Fig. 4.2 to take into account a wide pressure range in the formation of a vapour, the complete temperature-entropy chart for the vapour has the appearance shown in Fig. 4.3. Note its similarity with the temperature-enthalpy diagram dealt with in Level 3.

Once again the liquid line and the saturated vapour line are drawn which together enclose the wet vapour region. Again they join at the top of the critical point. Note, however, that the dry saturated vapour line is concave in this case, whereas it was convex in the case of the temperature-enthalpy diagram. On this chart, lines of constant dryness and of constant volume are often plotted in the wet region. The chart may be further extended by the introduction of a pressure axis, as shown. An important point here, however, is that this pressure axis will only correspond to pressures in the wet region. This point will be clear if reference is made to Fig. 4.3, where a constant pressure line is shown. Note that the line rises up to the transformation stage, is horizontal in the transformation stage and then rises up from this level into the superheat region. All along this line the pressure is constant. Since, however, the section in the transformation stage is horizontal, a vertical scale of pressure can be arranged to correspond with the pressures associated with these horizontal lines. To determine pressures in the

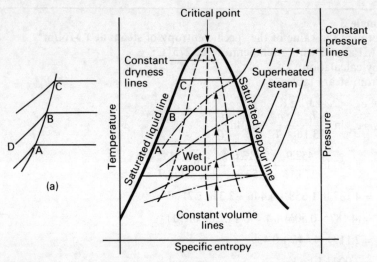

Fig. 4.3

superheat region and along the liquid line, the lines concerned must be traced along to the transformation stage and then the corresponding pressure determined from the vertical scale.

Note that the large area in the lower temperature region below the triple point is not usually shown in the temperature-entropy chart for steam since it has little value in the solution of problems concerning steam.

At (a) in Fig. 4.3 is shown an exaggerated section of the saturated liquid line, A, B, C. This shows the saturated liquid line as actually a locus through points A, B and C, while the liquid sections of the constant pressure lines below saturation temperature, such as DB, appear as separate lines. To normal scales, however, these separate liquid lines would near enough merge into a single line which is as illustrated in the main diagram of Fig. 4.3 as the saturated liquid line.

4.7 The isothermal process on the temperature-entropy chart

By definition an isothermal process is a process carried out at constant temperature.

Consider the vapour at some such state as at A in Fig. 4.4. Here the vapour is wet. Expansion from A will proceed as illustrated, the temperature remaining constant and, since it is in the wet region, the pressure will also remain constant until point B is reached, where the vapour becomes dry saturated. Further expansion takes the vapour into the superheat region shown as point C. In the superheat region the

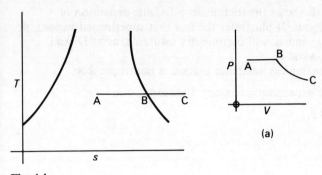

Fig. 4.4

pressure drops however. The pressure-volume diagram of the process is shown at (a). Since the vapour in the superheat region can be approximately considered to behave as a gas then the isothermal expansion BC could be considered as expanding according to the law $PV = C$, approximately.

4.8 The frictionless adiabatic process on the temperature-entropy chart – the isentropic process

An adiabatic process is defined as a process carried out such that there is no heat transferred during the process. The adiabatic therefore, neglecting friction, shock, etc., must have no area underneath it when plotted on a temperature-entropy diagram. It therefore appears as a vertical line showing it to have constant entropy.

A line of constant entropy is called an *isentropic*. The frictionless adiabatic process is therefore an isentropic process.

Process A–B shown in Fig. 4.5 shows the frictionless adiabatic expansion of very wet vapour. Note that at B the vapour has a higher dryness fraction than at A.

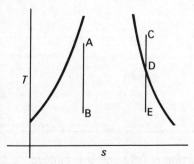

Fig. 4.5

Process CDE shows the frictionless adiabatic expansion of superheated vapour. It illustrates the fact that superheated vapour, if adiabatically expanded, will become dry saturated, as at D, and eventually wet, as at E.

Since the frictionless adiabatic process is isentropic then,

Entropy before expansion = Entropy after expansion

or, from Fig. 4.5,

$s_A = s_B$

and

$s_C = s_D = s_E$

Example 4
Steam at a pressure of 1 MN/m² and with a temperature of 250 °C is expanded adiabatically until its pressure becomes 0.2 MN/m². Determine the final dryness fraction of the steam. The adiabatic expansion may be assumed to be isentropic.

At 1 MN/m² and 250 °C the steam is superheated because saturation temperature at 1 MN/m² is 179.9 °C.

From tables, at 1 MN/m² and 250 °C,

$s_1 = 6.926$ kJ/kgK

For an isentropic process,

$s_1 = s_2$

and

$s_2 = s_{f2} + x s_{fg2}$

$= 1.530 + x_2(7.127 - 1.530)$

$= 1.530 + x_2 5.597$

$\therefore\ 6.926 = 1.530 + x_2 5.597$

from which

$x_2 = \dfrac{6.926 - 1.530}{5.597}$

$= \dfrac{5.396}{5.597}$

$x_2 = \underline{0.964}$

Example 5
Steam at 4 MN/m² has a dryness fraction of 0.92 and is expanded hyperbolically to 0.9 MN/m². It is then further expanded isentropically to 0.2 MN/m². Using steam tables, determine:

(a) the final condition of the steam;
(b) the overall change of specific entropy.

(a) For the hyperbolic expansion,

$$P_1 v_1 = P_2 v_2$$

From tables,

$$v_1 = x_1 v_{g1} = 0.92 \times 0.0498$$
$$= 0.0458 \text{ m}^3/\text{kg}$$
$$\therefore 4 \times 0.0458 = 0.9 \times v_2$$
$$v_2 = \frac{4 \times 0.0458}{0.9}$$
$$= 0.2036 \text{ m}^3/\text{kg}$$

At 0.9 MN/m², $v_{g2} = \underline{0.2149 \text{ m}^3/\text{kg}}$

∴ after hyperbolic expansion, the steam is wet.

$$x_2 = \frac{v_2}{v_{g2}} = \frac{0.2036}{0.2149} = \underline{0.947}$$

For an isentropic expansion

$$s_2 = s_3$$

and from tables, $s_{f2} = 2.094$ kJ/kgK

$$s_{fg2} = 4.529 \text{ kJ/kgK}$$

and

$$s_2 = s_{f2} + x_2 s_{fg2}$$
$$= 2.094 + 0.947 \times 4.529$$
$$= 2.094 + 4.289$$
$$= 6.383 \text{ kJ/kgK}$$
$$\therefore s_3 = \underline{6.383 \text{ kJ/kgK}}$$

At 0.2 MN/m², $s_g = 7.127$ kJ/kgK

∴ after isentropic expansion, the steam is wet

$$\therefore s_3 = s_{f3} + x_3 s_{fg3}$$
$$6.383 = 1.530 + (x_3 \times 5.597)$$
$$x_3 = \frac{6.383 - 1.530}{5.597}$$
$$= \frac{4.853}{5.597}$$
$$x_3 = \underline{0.867}$$

∴ the final condition of the steam at 0.2 MN/m^2 is wet with dryness fraction $= 0.867$

(b) $s_1 = s_{f1} + x_1 s_{fg1}$

$= 2.797 + 0.92 \times 3.273$

$= 2.797 + 3.011$

$= \underline{5.808 \text{ kJ/kgK}}$

∴ overall change of specific entropy

$= s_3 - s_1$

$= 6.383 - 5.808$

$= \underline{0.575 \text{ kJ/kgK}}$, an increase.

Example 6
7 kg of steam expands isentropically and adiabatically from a pressure of 5 MN/m^2 and temperature $350 \,°C$ to a pressure of 0.8 MN/m^2. The steam then expands to a pressure of 0.15 MN/m^2 and dryness fraction 0.97. This expansion appears as a straight line on a temperature–entropy chart. Determine:

(a) the overall heat transfer;
(b) the overall work done.

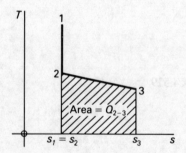

(a) For the isentropic process,

$s_1 = s_2$

From steam tables,

$s_1 = \underline{6.451 \text{ kJ/kgK}}$

∴ $s_2 = \underline{6.451 \text{ kJ/kgK}}$

Since this process is also adiabatic,

$Q_{1-2} = 0$

For state point 2.

$$s_2 = s_{f2} + x_2 s_{fg2}$$

$$\therefore 6.451 = 2.046 + (x_2 \times 4.617)$$

$$x_2 = \frac{6.451 - 2.046}{4.617}$$

$$= \frac{4.405}{4.617}$$

$$x_2 = \underline{0.954}$$

∴ steam is wet.

The temperature of this steam will be at saturation temperature for $0.8 \text{ MN/m}^2 = 170.4 \, °C$

$$\therefore T_2 = 170.4 + 273 = \underline{443.4 \text{ K}}$$

At state point 3, steam is wet and will be at saturation temperature for $0.15 \text{ MN/m}^2 = 111.4 \, °C$

$$\therefore T_3 = 111.4 + 273 = \underline{384.4 \text{ K}}$$

$$s_3 = s_{f3} + x_3 s_{fg3}$$

$$= 1.434 + (0.97 \times 5.789)$$

$$= 1.434 + 5.615$$

$$= \underline{7.049 \text{ kJ/kgK}}$$

Q_{2-3} = area of T–s diagram

$$= \frac{T_2 + T_3}{2}(s_3 - s_2)$$

$$= \frac{443.4 + 384.4}{2}(7.049 - 6.451)$$

$$= 413.9 \times 0.598$$

$$= \underline{247.51 \text{ kJ/kg}}$$

$$\therefore Q_{1-3} = Q_{1-2} + Q_{2-3}$$

$$= 0 + 247.51 \text{ kJ/kg}$$

$$= \underline{247.51 \text{ kJ/kg}}$$

∴ heat transfer for $7 \text{ kg} = 7 \times 247.51$

$$= \underline{1\,732.57 \text{ kJ}}$$

(b) For a non-flow process,
$$Q = \Delta u + W$$
∴ for the process 1–2,
$$0 = \Delta u + W$$
$$\therefore W_{1-2} = -\Delta u_{1-2}$$
Now $h = u + Pv$
$$\therefore u = h - Pv$$
$$\therefore u_1 = h_1 - P_1 v_1$$
$$= 3\,070 - 5 \times 10^3 \times 0.051\,9$$
$$= 3\,070 - 259.5$$
$$= \underline{2\,810.5 \text{ kJ/kg}}$$
$$u_2 = h_2 - P_2 v_2$$
$$h_2 = h_{f2} + h_{fg2}$$
$$= 721 + (0.954 \times 2\,048)$$
$$= 721 + 1\,953.8$$
$$= \underline{2\,674.8 \text{ kJ/kg}}$$
$$\therefore u_2 = 2\,674.8 - 0.8 \times 10^3 \times 0.954 \times 0.24$$
$$= 2\,674.8 - 183.17$$
$$= \underline{2\,491.63 \text{ kJ/kg}}$$
$$u_3 = h_3 - P_3 v_3$$
$$h_3 = 467 + (0.97 \times 2\,226)$$
$$= 467 + 2\,159.2$$
$$= \underline{2\,626.2 \text{ kJ/kg}}$$
$$\therefore u_3 = 2\,626.2 - 0.15 \times 10^3 \times 0.97 \times 1.159$$
$$= 2\,626.2 - 168.63$$
$$= \underline{2\,457.57 \text{ kJ/kg}}$$
∴ For process 1–2,
$$W_{1-2} = -\Delta u_{1-2}$$
$$= -(u_2 - u_1)$$
$$= -(2\,491.63 - 2\,810.5)$$
$$= \underline{318.87 \text{ kJ/kg}}$$

For process 2–3,

$$W_{2\text{-}3} = Q_{2\text{-}3} - \Delta u_{2\text{-}3} = Q_{2\text{-}3} - (u_3 - u_2)$$
$$= 247.51 - (2\,457.57 - 2\,491.63)$$
$$= 247.51 - (-34.06)$$
$$= 247.51 + 34.06$$
$$= \underline{281.57 \text{ kJ/kg}}$$

∴ overall work done

$$= W_{1\text{-}2} + W_{2\text{-}3}$$
$$= 318.87 + 281.57$$
$$= \underline{600.44 \text{ kJ/kg}}$$

∴ overall work done for 7 kg $= 7 \times 600.44$
$$= \underline{4\,203.08 \text{ kJ}}$$

4.9 The enthalpy-entropy chart for vapours

A commonly used vapour chart is the *enthalpy-entropy chart* as illustrated in Fig. 4.6.

Here the axes of enthalpy and entropy are used. The saturated

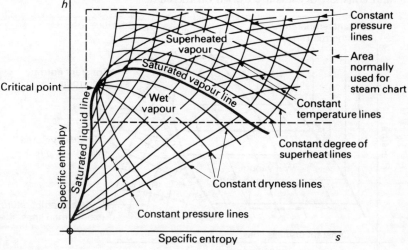

Fig. 4.6

liquid line and the saturated vapour line appear, meeting at the critical point. Lines of constant pressure cross the chart and lines of constant dryness fraction appear in the wet vapour region. In the superheated vapour region, constant temperature and constant degree of superheat lines are drawn.

This chart is very useful for the determination of property changes, such as change of enthalpy during an isentropic process which is illustrated as a vertical line on this chart.

A throttling process becomes a horizontal line on this chart.

Isothermal and constant pressure processes can be traced through on the chart.

In the case of the enthalpy-entropy chart for steam, it is usually only the area as shown dotted in Fig. 4.6 which is illustrated. This is because in steam plant, the steam turbine for example, it is only steam of good quality (high dryness fraction or superheated) which is useful and the expansions which occur can be easily plotted within the chart area illustrated.

This chart is sometimes called a Mollier chart.

4.10 The pressure-enthalpy chart for vapours

A chart which is often used for substances (refrigerants) used in the process of refrigeration is the *pressure-enthalpy chart*. Such substances as ammonia, methyl chloride and freon are used. Again, the saturated liquid and the saturated vapour lines appear and they join at the critical point as shown in Fig. 4.7. Lines of constant temperature (isothermal), constant entropy (isentropic), constant volume (isochoric) and constant dryness are drawn on the chart.

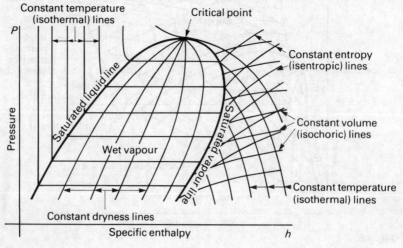

Fig. 4.7

This chart is particularly useful in the determination of property changes during constant pressure (isobaric) processes. The constant pressure process commonly occurs in refrigeration cycles and change of enthalpy during the constant pressure process is often required. This accounts for the choice of pressure and specific enthalpy as axes on a chart for refrigerants, (see Chapter 6).

4.11 The change of entropy for a gas (single-phase systems)

For a gas, the heat transferred during a non-flow process can be determined by using the non-flow energy equation,

$$dQ = dU + dW \tag{1}$$

Consider unit mass of gas and let its state change from pressure P_1, specific volume v_1 and temperature T_1 to new state P_2, v_2, T_2.

From equation (1),

$$dQ = c_v \, dT + P \, dv \tag{2}$$

Divide throughout by T, then,

$$\frac{dQ}{T} = \frac{c_v \, dT}{T} + \frac{P \, dv}{T} \tag{3}$$

But $dQ/T = ds$ and hence equation (3) now becomes,

$$ds = \frac{c_v \, dT}{T} + \frac{P \, dv}{T} \tag{4}$$

Now to obtain the change of entropy, this equation (4) must be integrated. However, it cannot be integrated as it stands because the expression $P \, dv/T$ contains too many variables. But this can be changed by using the characteristic equation, see *Thermodynamics Level 3*, section 4.3).

For unit mass of gas,

$$Pv = RT \text{ or } \frac{P}{T} = \frac{R}{v}$$

and substituting this in equation (4)

$$ds = c_v \frac{dT}{T} + R \frac{dv}{v} \tag{5}$$

Hence,

$$\int_{s_1}^{s_2} ds = c_v \int_{T_1}^{T_2} \frac{dT}{T} + R \int_{v_1}^{v_2} \frac{dv}{v}$$

Integrating,

$$\left[s\right]_{s_1}^{s_2} = c_v \left[\ln T\right]_{T_1}^{T_2} + R\left[\ln v\right]_{v_1}^{v_2}$$

or

$$s_2 - s_1 = c_v[\ln T_2 - \ln T_1] + R[\ln v_2 - \ln v_1]$$

or

$$s_2 - s_1 = c_v \ln \frac{T_2}{T_1} + R \ln \frac{v_2}{v_1} \qquad (6)$$

This equation (6) determines the change of specific entropy of a gas from a knowledge of temperature and volumes.

Now for a gas $c_p - c_v = R$ and substituting this into equation (6), then,

$$s_2 - s_1 = c_v \ln \frac{T_2}{T_1} + c_p \ln \frac{v_2}{v_1} - c_v \ln \frac{v_2}{v_1}$$

$$= c_p \ln \frac{v_2}{v_1} + c_v \left(\ln \frac{T_2}{T_1} - \ln \frac{v_2}{v_1}\right)$$

$$= c_p \ln \frac{v_2}{v_1} + c_v \ln \frac{T_2}{T_1} \frac{v_1}{v_2} \qquad (7)$$

Now from the characteristic equation,

$$\frac{P_1 v_1}{T_1} = \frac{P_2 v_2}{T_2}$$

from which,

$$\frac{T_2}{T_1} \frac{v_1}{v_2} = \frac{P_2}{P_1} \qquad (8)$$

Substituting equation (8) into equation (7), then,

$$s_2 - s_1 = c_p \ln \frac{v_2}{v_1} + c_v \ln \frac{P_2}{P_1} \qquad (9)$$

This equation (9) determines the change of specific entropy of a gas from a knowledge of volumes and pressures.

Again, from $c_p - c_v = R$, it follows that $c_v = c_p - R$ and substituting this into equation (6) then,

$$s_2 - s_1 = c_p \ln \frac{T_2}{T_1} - R \ln \frac{T_2}{T_1} + R \ln \frac{v_2}{v_1}$$

$$= c_p \ln \frac{T_2}{T_1} - R \left(\ln \frac{T_2}{T_1} - \ln \frac{v_2}{v_1}\right)$$

$$= c_p \ln \frac{T_2}{T_1} - R \ln \frac{T_2}{T_1} \frac{v_1}{v_2} \qquad (10)$$

Substituting equation (8) into equation (10) then,

$$s_2 - s_1 = c_p \ln \frac{T_2}{T_1} - R \ln \frac{P_2}{P_1} \tag{11}$$

This equation (11) determines the change of specific entropy of a gas from a knowledge of temperatures and pressures.

It should be noted that for any given change of state from P_1, v_1, T_1 to P_2, v_2, T_2 each of the equations (6), (9) and (11) will give the same result. The choice of equation is a matter of convenience only.

Change of entropy during a constant temperature (isothermal) process
Here $T_1 = T_2$ and hence $\ln T_2/T_1 = \ln 1 = 0$. Hence, from equation (6),

$$s_2 - s_1 = R \ln \frac{v_2}{v_1} \tag{12}$$

and, from equation (11),

$$s_2 - s_1 = -R \ln \frac{P_2}{P_1}$$

or

$$s_s - s_1 = R \ln \frac{P_1}{P_2} \tag{13}$$

From equation (9),

$$s_2 - s_1 = c_p \ln \frac{v_2}{v_1} + c_v \ln \frac{P_2}{P_1} \quad \text{(no change)}$$

Change of entropy during a constant volume (isochoric) process
Here $v_1 = v_2$, and hence $\ln v_2/v_1 = \ln 1 = 0$. Hence, from equation (6),

$$s_s - s_1 = c_v \ln \frac{T_2}{T_1} \tag{14}$$

and, from equation (9),

$$s_2 - s_1 = c_v \ln \frac{P_2}{P_1} \tag{15}$$

From equation (11),

$$s_2 - s_1 = c_p \ln \frac{T_2}{T_1} - R \ln \frac{P_2}{P_1} \quad \text{(no change)}$$

Change of entropy during a constant pressure (isobaric) process
Here, $P_1 = P_2$ and hence $\ln P_2/P_1 = \ln 1 = 0$. Hence, from equation (4),

$$s_2 - s_1 = c_p \ln \frac{v_2}{v_1} \tag{16}$$

and, from equation (11),

$$s_2 - s_1 = c_p \ln \frac{T_2}{T_1} \tag{17}$$

From equation (6)

$$s_2 - s_1 = c_v \ln \frac{T_2}{T_1} + R \ln \frac{v_2}{v_1} \quad \text{(no change)}$$

Change of entropy during the polytropic process $PV^n = C$
For the polytropic process it is shown that,

Heat transferred $= \dfrac{\gamma - n}{\gamma - 1} \times$ Work done (see *Thermodynamics Level 3*, section 5.12)

or, for unit mass of gas,

$$dQ = \frac{\gamma - n}{\gamma - 1} P \, dv \tag{18}$$

Divide equation (18) throughout by T, then,

$$\frac{dQ}{T} = ds = \frac{\gamma - n}{\gamma - 1} \frac{P \, dv}{T} \tag{19}$$

Again, for unit mass of gas,

$$Pv = RT \quad \text{or} \quad \frac{P}{T} = \frac{R}{v}$$

and substituting into equation (19)

$$ds = \frac{\gamma - n}{\gamma - 1} R \frac{dv}{v} \tag{20}$$

from this,

$$\int_{s_1}^{s_2} ds = \frac{\gamma - n}{\gamma - 1} R \int_{v_1}^{v_2} \frac{dv}{v}$$

Integrating,

$$\left[s \right]_{s_1}^{s_2} = \frac{\gamma - n}{\gamma - 1} R \left[\ln v \right]_{v_1}^{v_2}$$

or

$$s_2 - s_1 = \frac{\gamma - n}{\gamma - 1} R [\ln v_2 - \ln v_1]$$

or

$$s_2 - s_1 = \frac{\gamma - n}{\gamma - 1} R \ln \frac{v_2}{v_1} \tag{21}$$

Now $c_p - c_v = R$ and $c_p/c_v = \gamma$ or $c_p = \gamma c_v$; $\therefore \gamma c_v - c_v = R$ or $c_v(\gamma - 1) = R$ and substituting this into equation (21), then,

$$s_2 - s_1 = \frac{\gamma - n}{\gamma - 1} c_v(\gamma - 1) \ln \frac{v_2}{v_1}$$

or

$$s_2 - s_1 = c_v(\gamma - n) \ln \frac{v_2}{v_1} \qquad (22)$$

This expression gives the change of entropy in terms of volumes.
Now

$$\frac{v_2}{v_1} = \left(\frac{T_1}{T_2}\right)^{1/(n-1)} = \left(\frac{P_1}{P_2}\right)^{1/n} \qquad (23)$$

Substituting the temperature-volume relationship into equation (22), then

$$s_2 - s_1 = c_v(\gamma - n) \ln \left(\frac{T_1}{T_2}\right)^{1/(n-1)}$$

or

$$s_2 - s_1 = c_v \frac{(\gamma - n)}{(n - 1)} \ln \frac{T_1}{T_2} \qquad (24)$$

This expression gives the change of entropy in terms of temperatures.

Also, substituting the pressure-volume relationship from equation (23) into equation (22), then,

$$s_2 - s_1 = c_v(\gamma - n) \ln \left(\frac{P_1}{P_2}\right)^{1/n}$$

or

$$s_2 - s_1 = c_v \frac{(\gamma - n)}{n} \ln \frac{P_1}{P_2} \qquad (25)$$

This expression gives the change of entropy in terms of pressures.

It should again be noted that for a given polytropic change from P_1, v_1, T_1 to P_2, v_2, T_2, each of the equations (21), (22), (24) and (25) will give the same result. The choice is simply a matter of convenience.

It is important to remember that all the above expressions for the change of entropy are for unit mass of gas only. In any particular problem these expressions must be multiplied by the actual mass of gas being considered in order to determine the actual change of entropy which has occurred.

Note also, that in the above expressions, specific volume v has been used. Now for a mass m of the gas, its volume $V = mv$. The volume in the above expressions for the change of entropy appears in the ratio v_2/v_1.

Now

$$v_2/v_1 = mv_2/mv_1 = V_2/V_1$$

Thus, when calculating changes of entropy it is not always necessary to find the ratio of the specific volumes. The ratio of the actual volumes will give the same result.

4.12 The entropy chart for a gas

In order to make up a chart, numerical values of entropy must be known in order that points can be plotted on the chart. The entropy equations already determined are for changes of entropy and therefore do not give individual values at any particular state. The absolute value of entropy for a gas is not known because of the discontinuity of state as the gas is cooled. The gas liquifies and then solidifies before absolute zero of temperature is reached. However, it is the change of entropy which is important when discussing a process with a gas. This change of entropy is quite independent of any zero which may exist. Hence, when making up charts, it is usual to choose an arbitrary zero for entropy. Thus, for air, it is commonly arranged that its entropy is zero when its pressure is 0.101 MN/m² and its temperature is 0 °C.

Consider, now, the effect of this choice of zero on the equations for the change of entropy already determined.

Take equation (11), section 4.11, for example.

$$s_2 - s_1 = c_p \ln \frac{T_2}{T_1} - R \ln \frac{P_2}{P_1}$$

Assume that $s_1 = 0$, in which case, by the choice of the arbitrary zero $T_1 = 273.15$ K (say, 273 K), and $P_1 = 0.101$ MN/m².

Since $s_1 = 0$, then this equation becomes

$$s_2 = c_p \ln \frac{T_2}{273} - R \ln \frac{P_2}{0.101}$$

or, more generally, neglecting the suffix 2,

$$s = c_p \ln \frac{T}{273} - R \ln \frac{P}{0.101} \quad (P \text{ in MN/m}^2 \text{ in this case}) \tag{1}$$

Thus, from this equation (1) it is possible to determine the value of specific entropy s for a gas at any absolute temperature T and corresponding pressure P.

A similar arrangement is made for the other expressions. In the case of equations containing volume, the volume at the arbitrary zero condition is calculated, $= v_0$ say, and this is substituted in place of v_1. Thus in the case of equation (6), section 4.11, the value of entropy becomes

$$s = c_v \ln \frac{T}{273} + R \ln \frac{v}{v_0} \left(\text{specific volume } v_0 = R \frac{T_0}{P_0} \right) \quad (2)$$

Also, from equation (9)

$$s = c_p \ln \frac{v}{v_0} + c_v \ln \frac{P}{0.101} \quad (3)$$

Using these equations for the value of specific entropy it is now possible to make up a temperature-entropy chart.

The general appearance of the chart is illustrated in Fig. 4.8.

The chart usually has the axes of absolute temperature vertical and specific entropy horizontal. On the chart is drawn a network of constant volume and constant pressure lines, as shown. These lines can be drawn making use of the expressions already determined. Using equation (2) the constant volume lines can be drawn. By selecting a suitable constant volume and substituting into equation (2) the part of the expression $\ln(v/v_0)$ remains constant. Thus by selecting a number of values for the absolute temperature T then a series of values of s can be

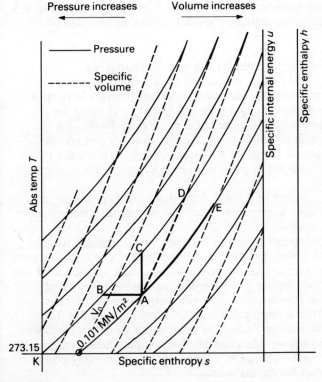

Fig. 4.8

determined for the constant volume v. These values can then be plotted to form a constant volume line. Taking various new values of v, then a series of constant volume lines can be drawn. A similar procedure is adopted when drawing the constant pressure lines. Here equation (3) can be used. Selecting a constant pressure P, then the part of the expression $\ln(P/0.101)$ now becomes constant. The procedure then follows that adopted for the constant volume lines.

The chart can have its usefulness extended by the introduction of the two vertical axes of specific internal energy and specific enthalpy.

When dealing with gas laws, it was shown that the internal energy of a gas is a function of temperature only (Joule's law). Thus the internal energy axis is parallel to the temperature axis. Now the change of specific internal energy for a gas $= c_v(T_2 - T_1)$. Assuming an arbitrary zero of internal energy of 0 °C, then the value of internal energy above 0 °C can be obtained from

$$u = c_v(T - 273)$$

Using this equation, the scale of specific internal energy can be introduced.

With regard to enthalpy, the change of specific enthalpy for a gas $= c_p(T_2 - T_1)$. Using a similar analogy as that used for the internal energy of a gas the value of the enthalpy above 0 °C can be obtained from,

$$h = c_p(T - 273)$$

Thus the scale of specific enthalpy can now be introduced.

Looking at Fig. 4.8, then, and selecting, say, point A, it will be observed that the constant volume and constant pressure lines cross, thus identifying the volume and pressure of the gas at this point. Further from the axes of the graph, it is possible to identify the temperature, specific internal energy, specific enthalpy and specific entropy of the gas for the point A. Thus a complete knowledge of the state of the gas is known at point A when it has been identified on the chart.

Again, reference to Fig. 4.8 will show that various processes, and the changes which occur during these processes, are readily determined on the chart.

An isothermal compression AB is shown as a horizontal straight line. It will be observed that the line AB finishes on a constant pressure line and the new pressure can be read from the value on the line. Point B does not coincide with a given constant volume line. The point B does, however, lie between two given constant volume lines. The volume at B is thus interpolated from these two given values. The change of specific entropy during the process can be determined from the specific entropy axis. The chart also shows that, during the isothermal process, no change in internal energy or enthalpy occurs since the temperature remains constant.

In the case of the frictionless adiabatic process, it must be remembered that, here, no heat is transferred during the process. Now the area under a process plotted on a temperature-entropy chart gives the heat transferred during the progress of the process. Since there is no heat transferred during the frictionless adiabatic process then there must be no area under the graph of the process when plotted on the temperature-entropy chart. For this to be so, the graph will appear as a vertical straight line which shows that the frictionless adiabatic process is carried out at constant entropy. A line of constant entropy is said to be an *isentropic* line and thus the frictionless adiabatic process is an *isentropic process*. The concept of the isentropic process as it applies to vapours is discussed in section 4.8.

Such a process as it applies to a gas is shown as AC in Fig. 4.8. This starts at point A and the new pressure, volume and temperature can be determined from point C. The change of specific internal energy and specific enthalpy can be determined from the axes of the chart.

A constant volume process from point A will appear as AD. The new pressure and temperature can be determined from the point D. Also, the change in specific internal energy, specific enthalpy and specific entropy can be determined from the axes.

A constant pressure process from point A will appear as AE. The new specific volume and temperature can be determined from the point E. Also, the change in specific internal energy, specific enthalpy and specific entropy can be determined from the axes.

It must be noted that the chart is usually made out for unit mass of gas, 1 kg of gas, for example. Thus, in any particular problem, the values obtained from the chart for the specific volume and changes in specific internal energy, specific enthalpy and specific entropy must be multiplied by the actual mass of gas being used.

Example 7

A quantity of gas has an initial pressure, volume and temperature of 2 MN/m^2, 0.1 m^3 and $300\,°\text{C}$. It is expanded to a pressure of 350 kN/m^3 according to the law $PV^{1.28} = C$. Determine:
(a) the change of entropy;
(b) the approximate change of entropy obtained by dividing the heat transferred by the gas by the mean absolute temperature during the expansion.

Take, $c_p = 1.04 \text{ kJ/kgK}$, $c_v = 0.74 \text{ kJ/kgK}$.

(a) For the gas,

$$R = c_p - c_v = 1.04 - 0.74 = 0.3 \text{ kJ/kgK}$$

and,

$P_1 V_1 = mRT_1$ (from the Characteristic Equation of a Perfect Gas)

$$\therefore m = \frac{P_1 V_1}{RT_1} = \frac{2 \times 10^3 \times 0.1}{0.3 \times 573} \qquad (T_1 = 300 + 273 = 573 \text{ K})$$

$$= \underline{1.163 \text{ kg}}$$

For 1 kg of gas,

$$s_2 - s_1 = c_p \ln \frac{V_2}{V_1} + c_v \ln \frac{P_2}{P_1}$$

Also,

$$P_1 V_2^n = P_2 V_2^n$$

$$\therefore V_2 = V_1 \left(\frac{P_1}{P_2}\right)^{1/n} = 0.1 \times \left(\frac{2}{0.35}\right)^{1/1.28} \quad \text{(pressures in MN/m}^2 \text{ in this case)}$$

$$= 0.1 \times 5.714^{1/1.28} = 0.1 \times 3.9$$

$$= \underline{0.39 \text{ m}^3}$$

$$\therefore s_2 - s_1 = 1.04 \ln \frac{0.39}{0.1} + 0.74 \ln \frac{0.35}{2}$$

$$= 1.04 \ln \frac{0.39}{0.1} - 0.74 \ln \frac{2}{0.35}$$

$$= 1.04 \ln 3.9 - 0.74 \ln 5.714$$

$$= (1.04 \times 1.361) - (0.74 \times 1.743)$$

$$= 1.416 - 1.29$$

$$= \underline{0.126 \text{ kJ/kgK}}$$

But the mass of gas $= 1.163$ kg.

$\therefore$ Change of entropy $= 0.126 \times 1.163$

$$= \underline{0.146 \ 5 \text{ kJ/K}}, \text{ an increase}$$

Note that the change of entropy could have been determined using the other expressions developed in the text. As examples:

$$s_2 - s_1 = c_p \ln \frac{T_2}{T_1} - R \ln \frac{P_2}{P_1}$$

Now, $\quad \dfrac{T_1}{T_2} = \left(\dfrac{P_1}{P_2}\right)^{(n-1)/n}$

$$\therefore T_2 = T_1 \left(\frac{P_2}{P_1}\right)^{(n-1)/n} = 573 \times \left(\frac{0.35}{2}\right)^{(1.28-1)/1.28}$$

$$= \frac{573}{5.714^{0.28/1.28}} = \frac{573}{5.714^{1/4.57}} = \frac{573}{1.464}$$

$$= \underline{391.4 \text{ K}}$$

or, $\quad \dfrac{T_1}{T_2} = \left(\dfrac{V_2}{V_1}\right)^{n-1}$

$\therefore T_2 = T_1 \left(\dfrac{V_1}{V_2}\right)^{n-1} = 573 \times \left(\dfrac{0.1}{0.39}\right)^{1.28-1} = \dfrac{573}{3.9^{0.28}}$

$$= \frac{573}{1.464}$$

$$= \underline{391.4 \text{ K}}$$

$\therefore s_2 - s_1 = 1.04 \ln \dfrac{391.4}{573} - 0.3 \ln \dfrac{0.35}{2}$

$$= -1.04 \ln \frac{573}{391.4} + 0.3 \ln \frac{2}{0.35}$$

$$= -1.04 \ln 1.464 + 0.3 \ln 5.714$$

$$= -(1.04 \times 0.381) + (0.3 \times 1.743)$$

$$= -0.396 + 0.522$$

$$= \underline{0.126 \text{ kJ/kgK}}$$

Alternatively,

$$s_2 - s_1 = c_v \ln \frac{T_2}{T_1} + R \ln \frac{V_2}{V_1}$$

$$= 0.74 \ln \frac{391.4}{573} + 0.3 \ln \frac{0.39}{0.1}$$

$$= -0.74 \ln \frac{573}{391.4} + 0.3 \ln \frac{0.39}{0.1}$$

$$= -0.74 \ln 1.464 + 0.3 \ln 3.9$$

$$= -(0.74 \times 0.381) + (0.3 \times 1.361)$$

$$= -0.282 + 0.408$$

$$= \underline{0.126 \text{ kJ/kgK}}$$

Alternatively,

$$s_2 - s_1 = c_v (\gamma - n) \ln \frac{V_2}{V_1}$$

and,

$$\gamma = \frac{c_p}{c_v} = \frac{1.04}{0.74} = 1.405$$

$$\therefore \; s_2 - s_1 = 0.74(1.405 - 1.28)\ln\frac{0.39}{0.1}$$

$$= 0.74 \times 0.125 \times 1.361$$

$$= \underline{0.126 \text{ kJ/kgK}}$$

Alternatively,

$$s_2 - s_1 = c_v \frac{(\gamma - n)}{(n-1)} \ln \frac{T_1}{T_2}$$

$$= 0.74 \frac{(1.405 - 1.28)}{(1.28 - 1)} \ln \frac{573}{391.4}$$

$$= 0.74 \times \frac{0.125}{0.28} \times 0.381$$

$$= \underline{0.126 \text{ kJ/kgK}}$$

Alternatively,

$$s_2 - s_1 = c_v \frac{(\gamma - n)}{n} \ln \frac{P_1}{P_2}$$

$$= 0.74 \frac{(1.405 - 1.28)}{1.28} \ln \frac{2}{0.35}$$

$$= 0.74 \times \frac{0.125}{1.28} \times 1.743$$

$$= \underline{0.126 \text{ kJ/kgK}}$$

(b) For a non-flow process,

$$Q = \Delta U + W$$

and for a polytropic process,

$$W = \frac{P_1 V_1 - P_2 V_2}{n - 1}$$

$$\therefore \; W = \frac{(2 \times 10^3 \times 0.1) - (350 \times 0.39)}{1.28 - 1}$$

$$= \frac{200 - 136.5}{0.28} = \underline{226.7 \text{ kJ}}$$

Alternatively,
$$W = \frac{mR(T_1 - T_2)}{n-1}$$
$$= \frac{1.163 \times 0.3(573 - 391.4)}{1.28 - 1}$$
$$= \frac{1.163 \times 0.3 \times 181.6}{0.28}$$
$$= \underline{226.3 \text{ kJ}} \quad \text{(Slight cumulative calculation difference)}$$

For a gas,
$$\Delta U = mc_v(T_2 - T_1)$$
$$= 1.163 \times 0.74(391.4 - 573)$$
$$= -1.163 \times 0.74 \times 181.6$$
$$= -\underline{156.3 \text{ kJ}}, \text{ a loss}$$
$$\therefore Q = -156.3 + 226.3$$
$$= \underline{70 \text{ kJ}}, \text{ a gain.}$$

Alternatively, for a polytropic process,
$$Q = \frac{(\gamma - n)}{(\gamma - 1)} \times \text{Work done}$$
$$= \frac{1.405 - 1.28}{1.405 - 1} \times 226.7$$
$$= \frac{0.125}{0.405} \times 226.7$$
$$= \underline{70 \text{ kJ}}, \text{ a gain.}$$

Mean absolute temperature $= \dfrac{T_1 + T_2}{2}$

$$= \frac{573 + 391.4}{2}$$
$$= \underline{482.2 \text{ K}}$$

$\therefore$ Approximate change of entropy $= \dfrac{70}{482.2}$

$$= \underline{0.145 \text{ kJ/K}}$$

Example 8

0.4 kg of air at a pressure of 1.5 MN/m² and temperature of 250 °C receives heat transferred while the pressure remains constant until the volume becomes 0.09 m³. Heat is then transferred from the gas while the volume remains constant while the pressure remains constant until the pressure becomes 550 kN/m². Determine the change of entropy during each process.

Take $c_p = 1.025$ kJ/kgK, $c_v = 0.727$ kJ/kgK.

For a gas,

$$R = c_p - c_v = 1.025 - 0.727 = \underline{0.298 \text{ kJ/kgK}}$$

and,

$$P_1 V_1 = mRT_1$$

$$\therefore V_1 = \frac{mRT_1}{P_1}$$

$$T_1 = 250 + 273 = \underline{523 \text{ K}}$$

$$\therefore V_1 = \frac{0.4 \times 0.298 \times 523}{1.5 \times 10^3} \quad \text{(Note } P_1 \text{ required in kN/m}^2 \text{ in this equation)}$$

$$= \underline{0.041\,6 \text{ m}^3}$$

For the constant pressure process,

$$\frac{V_1}{T_1} = \frac{V_2}{T_2}$$

$$\therefore T_2 = T_1 \frac{V_2}{V_1} = 523 \times \frac{0.09}{0.041\,6} = \underline{1\,131.5 \text{ K}}$$

Also,

$$s_2 - s_1 = c_p \ln \frac{V_2}{V_1} = 1.025 \ln \frac{0.09}{0.041\,6}$$

$$= 1.025 \ln 2.163\,5 = 1.025 \times 0.771\,7$$

$$= \underline{0.791 \text{ kJ/kgK}}, \text{ an increase}$$

$\therefore$ for 0.4 kg of air,

Change of entropy $= 0.4 \times 0.791$

$$= \underline{0.316\,4 \text{ kJ/K}}$$

Alternatively,

$$s_2 - s_1 = c_p \ln \frac{T_2}{T_1} = 1.025 \ln \frac{1\,131.5}{523}$$

$$= 1.025 \ln 2.163\,5 = 1.025 \times 0.771\,1$$
$$= \underline{0.791 \text{ kJ/kgK}}, \text{ as before.}$$

Alternatively,

$$s_2 - s_1 = c_v \ln \frac{T_2}{T_1} + R \ln \frac{V_2}{V_1}$$

$$= 0.727 \ln \frac{1\,131.5}{523} + 0.298 \ln \frac{0.09}{0.041\,6}$$

$$= 0.727 \ln 2.163\,5 + 0.298 \ln 2.163\,5$$

$$= (0.727 + 0.298) \ln 2.163\,5$$

$$= 1.025 \times 0.771\,7$$

$$= \underline{0.791 \text{ kJ/kgK}}, \text{ as before.}$$

For the constant volume process,

$$\frac{P_3}{T_3} = \frac{P_2}{T_2}$$

$$\therefore T_3 = T_2 \frac{P_3}{P_2} = 1\,131.5 \times \frac{550}{1.5 \times 10^3} = 414.9 \text{ K}$$

Also,

$$s_3 - s_2 = c_v \ln \frac{P_3}{P_2} = 0.727 \ln \frac{550}{1.5 \times 10^3}$$

$$= -0.727 \ln \frac{1.5 \times 10^3}{550} = -0.727 \ln 2.727$$

$$= -0.727 \times 1.003$$

$$= \underline{-0.729 \text{ kJ/kgK}}, \text{ a decrease.}$$

$\therefore$ For 0.04 kg of air,

Change of entropy $= 0.4 \times (-0.729)$

$$= \underline{-0.291\,6 \text{ kJ/K}}$$

Alternatively,

$$s_3 - s_2 = c_v \ln \frac{T_3}{T_2} = 0.727 \ln \frac{414.9}{1\,131.5}$$

$$= -0.727 \ln \frac{1\,131.5}{414.9} = -0.727 \ln 2.727$$

$= -0.727 \times 1.003$

$= \underline{-0.729 \text{ kJ/kgK}}$, as before.

Alternatively,

$$s_3 - s_2 = c_p \ln \frac{T_3}{T_2} - R \ln \frac{P_3}{P_2}$$

$$= 1.025 \ln \frac{414.9}{1\,131.5} - 0.298 \ln \frac{550}{1.5 \times 10^3}$$

$$= -1.025 \ln \frac{1\,131.5}{414.9} + 0.298 \ln \frac{1.5 \times 10^3}{550}$$

$$= -1.025 \ln 2.727 + 0.298 \ln 2.727$$

$$= (-1.025 + 0.298) \ln 2.727$$

$$= -0.727 \times 1.003$$

$$= \underline{-0.729 \text{ kJ/kgK}}, \text{ as before.}$$

Example 9

A quantity of gas at a pressure of 500 kN/m² occupies a volume of 0.7 m³ and has a temperature of 80 °C. The gas is compressed isothermally to a pressure of 2.4 MN/m². Determine the change of entropy.

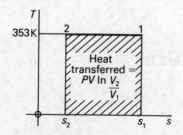

The solution of this problem relies on the original theory developed in section 4.1 that the area under a process plotted on a T–s chart is equal to the heat transferred reversibly during the process.

For a non-flow process, such as a compression,

$Q = \Delta U + W$

and for an isothermal process on a gas

$\Delta U = 0$

because the temperature remains constant.

Thus, for an isothermal process on a gas,

$$Q = W$$
and,
$$W = PV \ln r$$
where
$$r = \frac{V_2}{V_1} = \frac{P_1}{P_2}, \text{ since, } P_1 V_1 = P_2 V_2$$

$$\therefore W = 500 \times 0.7 \ln \frac{500}{2.4 \times 10^3} \qquad \text{(pressures in kN/m}^2\text{)}$$

$$= -\left(500 \times 0.7 \ln \frac{2.4 \times 10^3}{500}\right)$$

$$= -(500 \times 0.7 \ln 4.8)$$

$$= -(500 \times 0.7 \times 1.569)$$

$$= -\underline{549.15 \text{ kJ}}$$

$$= Q$$

Referring to the T–s chart shown in the diagram,

Area under the process plotted on the chart

$$= T_1(s_2 - s_1)$$
$$= 353(s_2 - s_1)$$
$$= Q$$

$$\therefore 353(s_2 - s_1) = -549.15$$

or,
$$s_2 - s_1 = -\frac{549.15}{353}$$
$$= -\underline{1.556 \text{ kJ/K}}, \text{ a decrease.}$$

4.13 Isentropic efficiency

Consider the steady-flow energy equation applied to the expansion of unit mass of substance through a turbine,

$$Z_1 + u_1 + P_1 v_1 + \frac{C_1^2}{2} + Q = Z_2 + u_2 + P_2 v_2 + \frac{C_2^2}{2} + W \qquad (1)$$

There will be little or no change in potential or kinetic energy through the turbine and hence the terms Z and $C^2/2$ can be neglected.

Although there is little or no change in kinetic energy the flow rate can be very high. This being the case there is very little time for heat exchange between the substance and the surroundings as the substance passes through the turbine.

Thus, theoretically $Q=0$, or, the expansion is theoretically adiabatic.

From this, then, equation (1) becomes,

$$u_1 + P_1 v_1 = u_2 + P_2 v_2 + W \tag{2}$$

from which,

$$W = (u_1 + P_1 v_1) - (u_2 + P_2 v_2)$$
$$= h_1 - h_2 \quad \text{(since } h = u + P_v = \text{specific enthalpy)} \tag{3}$$

Hence,

$$W = \text{change in specific enthalpy} \tag{4}$$

Now, as explained in section 4.8, if an adiabatic process is frictionless, the process is also isentropic, i.e., it is a process carried out at constant entropy.

Consider the adiabatic flow of steam through a steam turbine.

Figure 4.9 illustrates such a flow plotted on a T–s chart at (a) and an h–s chart at (b).

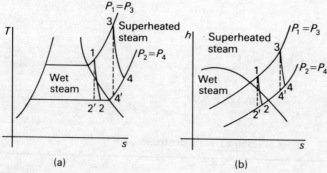

Fig. 4.9

Frictionless adiabatic, and hence isentropic expansions are shown as 1–2' and 3–4'. Note that expansion 1–2' starts as superheated steam at 1 and finishes as wet steam at 2'. Expansion 3–4' starts and finishes as superheated steam.

The pressure range in each case is from higher pressure $P_1 = P_3$ down to lower pressure $P_2 = P_4$. Now, in practice, with rapid expansion, internal friction is present which has the effect of making the enthalpy at the end of expansion higher than that which would have been the case had the expansion been frictionless. Referring to Fig. 4.9, the actual expansions, which include friction, are shown as 1–2 and 3–4.

In the case of expansion 1–2 in which the steam finishes wet, the condition 2 has a higher dryness fraction than condition 2′, or, $x_2 > x_{2'}$.

In the case of expansion 3–4 in which the steam is superheated at all times, the condition 4 has a higher temperature than the condition 4′, or, $T_4 > T_{4'}$.

In both cases, therefore, the enthalpy finishes at a higher value for the case including friction than the the theoretical case neglecting friction.

Refer now to equation (4), for unit mass,

W = change in specific enthalpy.

For the case neglecting friction,

$$W' = h_1 - h_{2'} \qquad (5)$$

For the case including friction,

$$W = h_1 - h_2$$

The ratio,

$$\frac{W}{W'} = \frac{h_1 - h_2}{h_1 - h_{2'}} \qquad (6)$$

is called the *isentropic efficiency*.

Note that, as expected, the effect of friction is that of reducing work output since $(h_1 - h_2) < (h_1 - h_{2'})$.

Further note, $(h_1 - h_2)$ is sometimes referred to as the *enthalpy drop*.

In the case of totally superheated expansion shown in Fig. 4.9(b)

Isentropic $\eta = \dfrac{h_3 - h_4}{h_3 - h_{4'}}$ (η = efficiency)

The concept of isentropic efficiency does not only apply to expansion of a substance through a steam turbine.

It will also apply to expansion through a steam nozzle, a gas nozzle, a gas turbine and a rotary type gas compressor.

Consider a nozzle which usually has a convergent–divergent cross-section as illustrated in Fig. 4.10.

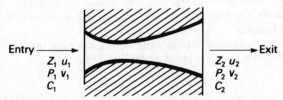

Fig. 4.10

The object of a nozzle is to expand a substance from a higher pressure P_1 to a lower pressure P_2 and accelerate the substance from a lower velocity C_1 to a higher velocity C_2.

Consider the steady-flow energy equation applied to the expansion of unit mass of substance through a nozzle.

The general form of the steady-flow energy equation is given in equation (1). There will be little or no change of potential energy through the nozzle and hence the terms Z can be neglected.

The flow rate of substance through the nozzle is very rapid and there is, therefore, little time for heat exchange between the substance and the surroundings. The expansion through the nozzle can, therefore, be considered as being adiabatic. Hence $Q=0$.

There is no external work done and hence $W=0$.

The steady-flow energy equation for flow through a nozzle becomes, therefore,

$$u_1 + P_1 v_1 + \frac{C_1^2}{2} = u_2 + P_2 v_2 + \frac{C_2^2}{2} \tag{7}$$

from which,

$$\frac{C_2^2 - C_1^2}{2} = (u_1 + P_1 v_1) - (u_2 + P_2 v_2)$$
$$= h_1 - h_2, \text{ since } h = u + Pv \tag{8}$$

Now, commonly, C_1 is small compared with C_2, and since the velocities are squared in equation (8), C_1 can be sensibly neglected. This being the case, equation (8) becomes,

$$\frac{C_2^2}{2} = h_1 - h_2 \tag{9}$$

or,

$$C_2 = \sqrt{\{2(h_1 - h_2)\}} \tag{10}$$

Note here in equation (10) the enthalpy drop $(h_1 - h_2)$ again appears.

Now, as with the case of the turbine already discussed, there will be internal friction which will occur as the substance expands through the nozzle.

Thus the enthalpy at exit from the nozzle will be higher than that which would occur had there been no friction. The enthalpy drop will thus be reduced and hence the exit velocity will be reduced as would be expected as a result of friction.

If the substance being expanded through the nozzle is steam then the expansion will appear on the T–s chart as in Fig. 4.9(a).

Again,

$$\text{Isentropic } \eta = \frac{h_1 - h_2}{h_1 - h_{2'}} \tag{11}$$

if the steam finishes wet, or,

Isentropic $\eta = \dfrac{h_3 - h_4}{h_3 - h_{4'}}$ (12)

if the steam is totally superheated.

In the cases of a gas turbine or rotary gas compressor the T–s chart will appear as in Fig. 4.11.

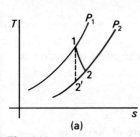 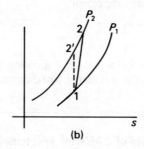

(a) (b)

Fig. 4.11

In Fig. 4.11(a) is illustrated the expansion in a gas turbine from higher pressure P_1 to lower pressure P_2. Again the enthalpy at the end of expansion will be higher due to friction and thus $h_2 > h_{2'}$.

Note, again, that the expansion 1–2′ is the frictionless adiabatic expansion which is also isentropic.

Expansion 1–2 is also an adiabatic expansion in that there is no heat transfer to the surroundings. However the expansion is not isentropic since $s_2 > s_1$.

Of theoretical interest it could be noted that if, during the expansion, an exact equivalent of the energy internally generated by friction could be transferred as heat to the surroundings, then the expansion would appear as 1–2′ which again would be isentropic. However, since heat has been transferred, the expansion is no longer adiabatic.

In the case of expansion through a gas turbine,

Isentropic $\eta = \dfrac{h_1 - h_2}{h_1 - h_{2'}}$ (13)

In Fig. 4.11(b) is illustrated the compression through a rotary gas compressor.

Frictionless adiabatic compression from lower pressure P_1 to higher pressure P_2 is shown as 1–2′ which is isentropic.

Due to friction, however, the actual adiabatic compression will appear as 1–2.

As a result of friction the final temperature will be higher and $T_2 > T_{2'}$.

The work of compression can be analysed as determined earlier in the case of expansion. Thus,

$$W = h_1 - h_2 \tag{14}$$

In this case however $h_1 < h_2$ and hence W is negative showing that work must be performed on a compressor for it to run. Hence,

$$-W = h_2 - h_1 \tag{15}$$

For the rotary gas compressor,

Isentropic $\eta = \dfrac{h_{2'} - h_1}{h_2 - h_1}$ \hfill (16)

Note that $(h_2 - h_1) > (h_{2'} - h_1)$ showing that the effect of friction is that of increasing the necessary work input.

Example 10
Steam at a pressure of 2 MN/m² and temperature of 300 °C is expanded through a turbine to a pressure of 0.2 MN/m² with an isentropic efficiency of 0.87. The flow rate of the steam through the turbine is 10 kg/s. Determine:
(a) the final quality of the steam;
(b) the power output from the turbine in kilowatts.

(a) Using steam tables, at 2 MN/m² and 300 °C

$$h_1 = 3\,025 \text{ kJ/kg}$$
$$s_1 = 6.768 \text{ kJ/kgK}$$

For frictionless adiabatic expansion,

$$s_1 = s_{2'} = 6.768 \text{ kJ/kgK}$$

At 0.2 MN/m²,

$$s_g = 7.127 \text{ kJ/kgK}$$

∴ the steam after expansion is wet.

$$s_{2'} = s_{f2} + x_{2'}(s_{fg})$$
$$= 1.530 + x_{2'}(5.597) = 6.768$$
$$x_{2'} = \frac{6.768 - 1.530}{5.597} = \frac{5.238}{5.597} = \underline{0.936}$$
$$h_{2'} = h_{f2} + x_{2'}(h_{fg})$$
$$= 505 + 0.936(2\,202)$$
$$= 505 + 2\,061.07$$
$$= \underline{2\,566.07 \text{ kJ/kg}}$$

Isentropic $\eta = \dfrac{h_1 - h_2}{h_1 - h_{2'}}$

$\therefore h_2 = h_1 - (h_1 - h_{2'})\eta$

$= 3\,025 - (3\,025 - 2\,566.07)0.87$

$= 3\,025 - (458.93)0.87$

$= 3\,025 - 399.27$

$= \underline{2\,625.73 \text{ kJ/kg}}$

$h_2 = h_{f2} + x_2(h_{fg})$

$\therefore x_2 = \dfrac{h_2 - h_{f2}}{h_{fg}}$

$= \dfrac{2\,625.73 - 505}{2\,202}$

$= \dfrac{2\,120.73}{2\,202}$

$= \underline{0.963} = $ final quality of the steam.

(b) $\quad W = h_1 - h_2$

$= 3\,025 - 2\,625.73$

$= \underline{399.27 \text{ kJ/kg}}$

For a steam flow rate of 10 kg/s

$\dot{W} = 399.27 \times 10$

$= 3\,992.7 \text{ kJ/s}$

$= \underline{3\,992.7 \text{ kW}} \quad$ (since $1 \text{ kJ/s} = 1 \text{ kW}$)

= output from the turbine.

Note: This expansion is illustrated on steam T–s and h–s charts as 1–2 in Fig. 4.9.

Example 11
Steam enters a nozzle at a pressure of 1 MN/m^2 and with a temperature of 250 °C. The steam expands through the nozzle to a pressure of 0.15 MN/m^2 with an isentropic efficiency of 0.95. The flow rate of the steam is 2.5 kg/s. Determine:
(a) the final dryness of the steam at exit from the nozzle;
(b) the exit velocity of the steam from the nozzle in m/s;
(c) the exit area of the nozzle in mm^2.

(a) From steam tables,
$$h_1 = \underline{2\,944\text{ kJ/kg}}$$
$$s_1 = \underline{6.926\text{ kJ/kgK}}$$
For frictionless adiabatic expansion,
$$s_1 = s_{2'} = \underline{6.926\text{ kJ/kgK}}$$
At 0.15 MN/m^2
$$s_g = \underline{7.223\text{ kJ/kgK}}$$
$\therefore$ steam after expansion is wet.
$$s_{2'} = s_{f2} + x_{2'}(s_{fg})$$
$$\therefore 6.926 = 1.434 + x_{2'}(5.789)$$

from which,
$$x_{2'} = \frac{6.926 - 1.434}{5.789} = \frac{5.492}{5.789}$$
$$= \underline{0.949}$$
$$h_{2'} = h_{f2} + x_{2'}(h_{fg2})$$
$$= 467 + 0.949(2\,226) = 467 + 2\,112.47$$
$$= \underline{2\,579.47\text{ kJ/kg}}$$
Isentropic $\eta = \dfrac{h_1 - h_2}{h_1 - h_{2'}}$

from which,
$$h_1 - h_2 = (h_1 - h_{2'})\eta = 0.95(2\,944 - 2\,579.47)$$
$$= 0.95 \times 364.53$$
$$= \underline{346.3\text{ kJ/kg}}$$
$$\therefore h_2 = 2\,944 - 346.3$$
$$= \underline{2\,597.7\text{ kJ/kg}}$$
$$h_2 = h_{f2} + x_2(h_{fg2})$$
$$\therefore 2\,597.7 = 467 + x_2(2\,226)$$

from which,
$$x_2 = \frac{2\,597.7 - 467}{2\,226}$$
$$= \underline{0.957} = \text{final dryness of the steam.}$$

(b)　$C_2 = \sqrt{\{2(h_1 - h_2)\}}$
　　　　$= \sqrt{(2 \times 346.3 \times 10^3)}$　　(Note h in J/kg in this equation)
　　　　$= \sqrt{692\,600}$
　　　　$= \underline{832.23 \text{ m/s}}$

(c)　$A_2 = \dfrac{\dot{m}v_2}{C_2}$　where $\dot{m}$ = flow rate kg/s
　　　　　　　　　　　　v_2 = specific volume of steam at exit, m³/kg
　　　　　　　　　　　　C_2 = exit velocity m/s
　　　　　　　　　　　　A_2 = exit area, m²

　　$v_2 = x_2 v_{g2} = 0.957 \times 1.159$
　　　　$= \underline{1.109 \text{ m}^3/\text{kg}}$

　$\therefore A_2 = \dfrac{2 \times 1.109}{832.23}$

　　　　$= \underline{0.002\,665 \text{ m}^2}$
　　　　$= \underline{2\,665 \text{ mm}^2}$

Example 12
Air enters a gas turbine at a pressure of 600 kN/m² and with a temperature of 800 °C. The air expands in the turbine to an exhaust pressure of 116.5 kN/m². The isentropic efficiency of the expansion is 87%. The mass flow rate of the air through the turbine is 2.5 kg/s.

For the air, take $c_p = 1.005$ kJ/kgK; $\gamma = 1.4$.
Determine:
(a) the exhaust air temperature as it leaves the turbine;
(b) the power output from the turbine in kilowatts.

(a) This type of expansion is illustrated on a T–s chart in Fig. 4.11(a).

For a gas,

$$\frac{T_1}{T_{2'}} = \left(\frac{P_1}{P_2}\right)^{(\gamma - 1)/\gamma}$$

$\therefore T_{2'} = T_1 \left(\dfrac{P_2}{P_1}\right)^{(\gamma - 1)/\gamma}$　and $T_1 = 800 + 273 = \underline{1\,073 \text{ K}}$

　　　$= 1\,073 \times \left(\dfrac{116.5}{600}\right)^{(1.4 - 1)/1.4} = 1\,073 \times \left(\dfrac{1}{5.15}\right)^{0.4/1.4}$

　　　$= \dfrac{1\,073}{5.15^{1/3.5}} = \dfrac{1\,073}{1.597}$

　　　$= \underline{671.9 \text{ K}}$

$= 671.9 - 273$

$= \underline{398.9\,°C} =$ exhaust air temperature.

(b) Isentropic $\eta = \dfrac{h_1 - h_2}{h_1 - h_{2'}}$ and for a gas, $h_1 - h_2 = c_p(T_1 - T_2)$

∴ for a gas,

Isentropic $\eta = \dfrac{c_p(T_1 - T_2)}{c_p(T_1 - T_{2'})} = \dfrac{T_1 - T_2}{T_1 - T_{2'}}$

∴ in this case,

$$0.87 = \dfrac{1\,073 - T_2}{1\,073 - 671.9}$$

from which,

$T_2 = 1\,073 - 0.87(1\,073 - 671.9)$

$= 1\,073 - 0.87(401.1)$

$= 1\,073 - 348.96$

$= \underline{724.04\text{ K}}$

$= 724.04 - 273$

$= \underline{451.04\,°C}$

Power output $= \dot{m}(h_1 - h_2)$

$= \dot{m}c_p(T_1 - T_2)$

$= 2.5 \times 1.005 \times (1\,073 - 724.04)$

$= 2.5 \times 1.005 \times 348.96$

$= \underline{876.76\text{ kW}}$

Example 13

A rotary air compressor takes in air at a pressure of 110 kN/m^2 and with a temperature of $20\,°C$. The air is compressed to a pressure of 500 kN/m^2. The isentropic efficiency of compression is 84%. The flow rate of the air is 3 kg/s.

For the air, take $c_p = 1.006$ kJ/kgK;
$c_v = 0.717$ kJ/kgK.

Determine:
(a) the air temperature after compression;
(b) the power input required to drive the compressor.

(a) This type of compression is illustrated on a T–s chart in Fig. 4.11(b).

For a gas,

$$\gamma = \frac{c_p}{c_v} = \frac{1.006}{0.717} = \underline{1.403}$$

$$\frac{T_1}{T_{2'}} = \left(\frac{P_1}{P_2}\right)^{(\gamma-1)/\gamma}$$

$$\therefore \quad T_{2'} = T_1\left(\frac{P_2}{P_1}\right)^{(\gamma-1)/\gamma} \text{ and } T_1 = 20 + 273 = \underline{293 \text{ K}}$$

$$= 293 \times \left(\frac{500}{110}\right)^{(1.403-1)/1.403} = 293 \times 4.545^{0.403/1.403}$$

$$= 293 \times 1.545$$

$$= \underline{452.69 \text{ K}}$$

$$= 452.69 - 273$$

$$= \underline{179.69\,°\text{C}}$$

For a rotary air compressor.

Isentropic $\quad \eta = \dfrac{h_{2'} - h_1}{h_2 - h_1}$ and for a gas, $h_2 - h_1 = c_p(T_2 - T_1)$

$$= \frac{c_p(T_{2'} - T_1)}{c_p(T_2 - T_1)} = \frac{T_{2'} - T_1}{T_2 - T_1}$$

$\therefore$ in this case,

$$0.84 = \frac{452.69 - 293}{T_2 - 293}$$

from which,

$$T_2 = \frac{(T_{2'} - T_1)}{\eta} + T_1$$

$$= \frac{(452.69 - 293)}{0.84} + 293$$

$$= \frac{159}{0.84} + 293 = 190.11 + 293$$

$$= \underline{483.11 \text{ K}}$$

$$= 483.11 - 273$$

$$= \underline{210.11\,°\text{C}} = \text{air temperature after compression.}$$

(b) $W = \dot{m}(h_1 - h_2) = \dot{m}c_p(T_1 - T_2)$

$\quad = -\dot{m}c_p(T_2 - T_1)$

$\quad = -\{3 \times 1.006 \times (483.11 - 293)\}$

$\quad = -\{3 \times 1.006 \times 190.11\}$

$\quad = -573.75 \text{ kW}$

Note that the negative sign shows that work must be done on the compressor.

Questions 4

1 Steam initially at a pressure of 3 MN/m² and with a temperature of 350 °C is expanded reversibly to a pressure of 0.16 MN/m² and a dryness fraction of 0.95. The expansion appears as a straight line when plotted on a temperature–entropy chart. After the expansion, the steam is then cooled at constant temperature until the dryness fraction becomes 0.6. Determine:
(a) the heat transferred/kg steam during each process;
(b) the work transfer/kg steam during each process.

[(a) 86.14 kJ/kg, received; 777.04 kJ/kg, rejected; (b) 512.94 kJ/kg; -60.94 kJ/kg]

2 Steam at a pressure of 40 bar and temperature 300 °C is expanded to a pressure of 8 bar according to the law $PV^{1.22}$ = constant. For this expansion, determine:
(a) the final condition of the steam;
(b) the heat transfer/kg steam;
(c) the change of specific entropy.

[(a) Steam wet, dryness fraction 0.915; (b) -39.39 kJ/kg, a loss; (c) -0.093 kJ/kgK, a decrease]

3 Steam at a pressure of 2 MN/m² has a dryness fraction of 0.9 and occupies a volume of 0.5 m³. It is expanded hyperbolically (PV = constant) until the pressure becomes 0.6 MN/m². The steam is then expanded isentropically until the pressure becomes 0.2 MN/m². Determine:
(a) the change of entropy during the hyperbolic process;
(b) the final condition of the steam.
[(a) 3.066 2 kJ/K, an increase; (b) steam wet, dryness fraction 0.89]

4 Steam at a pressure of 5 MN/m² and with a temperature of 350 °C is expanded isentropically to a pressure of 1 MN/m². The steam is

then reheated while the pressure remains constant at 1 MN/m² until the temperature becomes 250 °C. It is then further expanded isentropically to a pressure of 0.2 MN/m². Determine:
(a) the condition of the steam after each isentropic expansion;
(b) the heat transfer required/kg steam to carry out the constant pressure reheat.

[(a) Steam wet in each case, dryness fractions 0.97, 0.964;
(b) 316.3 kJ/kg, received]

5 Steam at a pressure of 1.6 MN/m² is throttled to a pressure of 0.15 MN/m² and a temperature of 150 °C. The steam is then expanded to a pressure of 50 kN/m² according to the law $PV^{1.27}$ = constant. Determine:
(a) the original condition of the steam before being throttled;
(b) the final condition of the steam after the expansion;
(c) the change of entropy during each process.

[(a) Steam wet, dryness fraction 0.989; (b) steam wet, dryness fraction 0.943; (c) 1.043 kJ/kgK, an increase; 0.198 kJ/kgK, an increase.]

6. A quantity of gas has an initial pressure, volume and temperature of 850 kN/m², 0.65 m³ and 170 °C, respectively. It is expanded to a pressure of 150 kN/m² according to the law $PV^{1.34}$ = constant. Determine the change of entropy and state whether it is an increase or a decrease. Take $c_p = 0.84$ kJ/kgK, $c_v = 0.59$ kJ/kgK.

[0.468 kJ/K, an increase.]

7. A turbine is supplied with steam at a pressure of 3 MN/m² and a temperature of 350 °C. The steam is expanded through the turbine to a pressure of 0.4 MN/m² and a dryness fraction of 0.985. The power output from the turbine is 2 megawatts. Determine:
(a) the isentropic efficiency of the turbine;
(b) the flow rate of the steam through the turbine in kg/s.

[(a) 92.8%; 4.88 kg/s.]

8 Air enters a gas turbine at a pressure of 700 kN/m² and with a temperature of 900 °C. The air leaves the turbine at a pressure of 175 kN/m² and with a temperature of 560 °C. The power output of the turbine is 650 kW.
Take, $c_p = 1.007$ kJ/kgK; $\gamma = 1.397$. Determine:
(a) the isentropic efficiency of the gas turbine;
(b) the mass flow rate of the air in kg/s.

[(a) 89%; (b) 1.898 kg/s.]

9 A gas enters a rotary compressor at a pressure of 150 kN/m² and with a temperature of 30 °C. The gas is compressed through a pressure ratio of 4.5:1 with an isentropic efficiency of 82%. The

power input to the compressor is 200 kW.
Take, $c_v = 0.75$ kJ/kgK; $\gamma = 1.404$. Determine:
(a) the gas pressure and temperature after compression;
(b) the mass flow rate of the gas in kg/s.

[(a) 675 kN/m^2; 230.3 °C; (b) 0.948 kg/s.]

10 Steam enters a nozzle at a pressure of 1.5 MN/m^2 and with a temperature of 250 °C. It expands through the nozzle to a pressure of 300 kN/m^2 with an isentropic efficiency of 90%. The exit area of the nozzle is 1 500 mm^2. Determine:
(a) the final quality of the steam;
(b) the exit velocity of the steam from the nozzle in m/s;
(c) the mass flow rate of the steam in kg/s.

[(a) steam wet, dryness fraction 0.956; (b) 578.15 m/s; 1.498 kg/s.]

11 A gas enters a nozzle at a pressure of 3 MN/m^2 and with a temperature of 400 °C. It leaves with a pressure of 0.75 MN/m^2. The expansion through the nozzle has an isentropic efficiency of 88%. The mass flow rate of the gas through the nozzle is 2 kg/s.
Take $c_p = 1.04$ kJ/kgK; $c_v = 0.75$ kJ/kgK. Determine:
(a) the temperature of the gas as it leaves the nozzle;
(b) the velocity of the gas as it leaves the nozzle.

[(a) 210.1 °C; 628.48 m/s.]

Chapter 5

Gas and vapour power cycles

5.1

If a substance passes through a series of processes such that it is eventually returned to its original state then the substance is said to have been taken through a cycle (see section 1.2 and Level 3 Thermodynamics).

During a cycle there will be some heat transfer and some work transfer to and from the substance.

Since after performing a cycle, the substance is returned to its original state, then, by the First Law of Thermodynamics,

$$\oint Q = \oint W, \quad \text{(see section 1.2)} \tag{1}$$

Thus, for a cycle, the net work transfer can be determined by an analysis of the net heat transfer, or,

Work done = Heat received − Heat rejected (2)

The ratio, $\dfrac{\text{Work done}}{\text{Heat received}}$

is called the *thermal efficiency* (see Level 3), or

Thermal $\eta = \dfrac{W}{Q}$, (3)

where, W = Work done, Q = Heat received

Note also that, because the area under a process illustrated on a pressure–volume graph is equal to the work done, then, the net area of a pressure–volume diagram of a cycle is equal to the net work of the cycle.

This, therefore, gives another method by which the net work of a cycle can be determined.

The equation,

$$\text{Thermal } \eta = \frac{W}{Q}$$

gives the *theoretical* or *ideal* thermal efficiency. The *actual* thermal efficiency of a practical cycle is given by the equation,

$$\text{Actual thermal } \eta = \frac{\text{actual work done}}{\text{thermal energy from fuel}} \quad (4)$$

This is always less than the theoretical thermal efficiency.

A practical cycle is carried out in an engine or turbine and will incur many losses which will include heat transfer loss, fuel combustion loss, non-uniform energy distribution in the working substance, friction, leakage, the need to keep temperatures within practical working limits, the running of auxiliary equipment such as pumps, alternators, valve gear and cooling equipment.

The ratio between the actual thermal efficiency and the ideal thermal efficiency is called the *relative efficiency* or *efficiency ratio*, thus,

$$\text{Relative efficiency} = \frac{\text{actual thermal efficiency}}{\text{ideal thermal efficiency}} \quad (5)$$

Another useful concept in the consideration of a cycle is that of the *work ratio*. This is defined as:

$$\text{Work ratio} = \frac{\text{net work done}}{\text{positive work done}} \quad (6)$$

where, net work done = positive work done − negative work done.

Note that, from equation (6), if the negative work is reduced then the work ratio $\to 1$.

A cycle with good ideal thermal efficiency together with a good work ratio suggests good overall efficiency potential in a practical power producing plant using the cycle.

Work ratio can give a comparative indication of plant size. A plant with a low work ratio would suggest that the work components of the plant are larger when compared with a plant which has a higher work ratio and similar power output.

Work ratio is most commonly applied to such cycles as arranged in steam plant and gas turbines. Such plants are composed of separate units each performing a particular function. In a steam plant there is the boiler, the engine or turbine, the condenser and the feed pump (see section 5.8). In a gas turbine there is the compressor, the combustion chamber (or chambers) and the turbine (see section 5.4).

Another form of comparison between cycles is by means of the *specific steam or fuel consumption*.

In the case of steam plant,

$$\text{Specific steam consumption} = \frac{\text{mass flow of steam in kg/h}}{\text{power output in kW}} \quad (7)$$

This gives the mass of steam used per unit power output in kg/kWh.

In the case of internal combustion engines such as the gas turbine and the petrol or diesel engine the specific fuel consumption is used where,

$$\text{Specific fuel consumption} = \frac{\text{mass of fuel used in kg/h}}{\text{power output in kW}} \quad (8)$$

This gives the mass of fuel used per unit power output in kg/kWh.

Thus a cycle which has a lower specific steam or fuel consumption indicates that it has better energy conversion performance than a cycle with a higher specific steam or fuel consumption.

The specific steam or fuel consumption can also be determined from a knowledge of the specific work output.

If

$$\text{Specific work output} = W_s \frac{\text{kJ}}{\text{kg (steam or fuel)}}$$

then,

$$\frac{1}{W_s}\frac{\text{kg}}{\text{kJ}} = \frac{1}{W_s}\frac{\text{kg}}{\text{kWs}} = \frac{3600}{W_s}\frac{\text{kg}}{\text{kWh}}$$

$$= \text{specific steam or fuel consumption} \quad (9)$$

$$\left(\text{Note that } 1\,\text{kW} = \frac{1\,\text{kJ}}{\text{s}} \quad \therefore \quad 1\,\text{kJ} = 1\,\text{kWs}\right)$$

In the case of reciprocating engines such as steam, petrol and diesel engines which use a piston–crank mechanism, a means of comparison between cycles can be made by means of the *mean effective pressure*. This is that the theoretical pressure which, if it was maintained constant throughout the volume change of the cycle (engine stroke of a practical cycle), would give the same work output as that obtained from the cycle.

Figure 5.1 illustrates a cycle plotted on a P–V diagram and operating between the volume limits of V_1 and V_2.

The area of the cycle diagram will determine the work done $= W$

This is called the *indicated* work done.

The stroke volume of the diagram $= V_1 - V_2$.

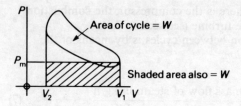

Fig. 5.1

The mean effective pressure is determined by the equation,

$$P_M = \frac{W}{V_1 - V_2}, \quad \text{where } P_M = \text{mean effective pressure} \tag{9}$$

Note that $W/(V_1 - V_2)$ is also the work done/unit swept volume.

Thus a cycle with a higher mean effective pressure will indicate that it has better work characteristic than a cycle with a lower mean effective pressure.

A further note concerning the cycles used in internal combustion engines is that, if the effects of the fuel used is neglected, the gas can be considered to closely approximate to that of air alone. Thus, the theoretical cycles such as the constant volume, constant pressure and the Diesel cycles are sometimes referred to as the *air standard cycles* and the related efficiencies are referred to as *air standard efficiencies.*

5.2 The Carnot cycle for a gas

Carnot, (see section 2.3), conceived a cycle made of thermodynamically reversible processes. By calculating the thermal efficiency of this cycle it is possible to establish the maximum possible efficiency between the temperature limits taken. Figure 5.2(a) shows the Carnot cycle illustrated on a $P-V$ diagram. It will be observed that it is made up of

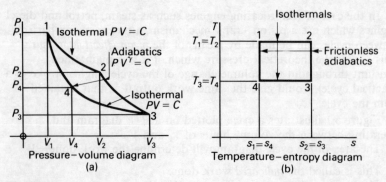

Fig. 5.2

four reversible processes so put together that they form a closed cycle. Thus by proceeding round the cycle it is possible to return to the original state and hence the cycle admits of repetition.

The processes are as follows:

1–2 Isothermal expansion.

Pressure falls from P_1 to P_2.
Volume increases from V_1 to V_2.
Temperature remains constant at $T_1 = T_2$.

$$\text{Work done} = P_1 V_1 \ln \frac{V_2}{V_1} = mRT_1 \ln \frac{V_2}{V_1}$$

For the isothermal, $Q = W$,

$$\therefore \text{ Heat received} = mRT_1 \ln \frac{V_2}{V_1}$$

2–3 Adiabatic expansion.

Pressure falls from P_2 to P_3.
Volume increases from V_2 to V_3.
Temperature falls from T_2 to T_3.

$$\text{Work done} = \frac{P_2 V_2 - P_3 V_3}{\gamma - 1} = mR(T_2 - T_3)^{\gamma - 1}$$

For the adiabatic, $Q = 0$.

$\therefore$ no heat transfer during this process

3–4 Isothermal compression.

Pressure increases from P_2 to P_4.
Volume reduced from V_3 to V_4.
Temperature remains constant $T_3 = T_4$.

$$\text{Work done} = P_3 V_3 \ln \frac{V_4}{V_3} = -P_3 V_3 \ln \frac{V_3}{V_4}$$

$$= -mRT_3 \ln \frac{V_3}{V_4}$$

For the isothermal, $Q = W$,

$$\therefore \text{ Heat rejected} = mRT_3 \ln \frac{V_3}{V_4}$$

4–1 Adiabatic compression.

Pressure increases from P_4 to P_1.
Volume reduced from V_4 to V_1.
Temperature increases from T_4 to T_1.

$$\text{Work done} = \frac{P_4V_4 - P_1V_1}{\gamma - 1} = -\frac{(P_1V_1 - P_4V_4)}{\gamma - 1}$$

$$= -\frac{mR(T_1 - T_4)}{\gamma - 1}$$

For the adiabatic $Q = 0$.

∴ no heat transfer during this process.

Note that this process returns the gas to its original state at 1.

The work done during this cycle may be determined by summing the areas beneath the various processes taking the expansions as positive areas and compressions as negative areas. Thus,

Work done/cycle = area under 1–2 + area under 2–3
 − area under 3–4 − area under 4–1
 = area 1 2 3 4
 = area enclosed by cycle

or

$$\text{Work done/cycle} = mRT_1 \ln \frac{V_2}{V_1} + \frac{mR(T_2 - T_3)}{\gamma - 1}$$

$$- mRT_3 \ln \frac{V_3}{V_4} - \frac{mR(T_1 - T_4)}{\gamma - 1} \qquad (1)$$

Now $T_1 = T_2$ and $T_3 = T_4$, from the isothermals.

$$\therefore \frac{mR(T_2 - T_3)}{\gamma - 1} = \frac{mR(T_1 - T_4)}{\gamma - 1}$$

Hence, from equation (1),

$$\text{Work done/cycle} = mRT_1 \ln \frac{V_2}{V_1} - mRT_3 \ln \frac{V_3}{V_4} \qquad (2)$$

Now for the adiabatic 1–4,

$$\frac{T_1}{T_4} = \left(\frac{V_4}{V_1}\right)^{\gamma - 1} \qquad (3)$$

for the adiabatic 2–3,

$$\frac{T_2}{T_3} = \left(\frac{V_3}{V_2}\right)^{\gamma - 1} \qquad (4)$$

But $T_1 = T_2$ and $T_3 = T_4$,

$$\therefore \frac{T_1}{T_4} = \frac{T_2}{T_3} \qquad (5)$$

Hence, from equations (3) and (4)

$$\frac{V_4}{V_1} = \frac{V_3}{V_2} \quad \text{or} \quad \frac{V_2}{V_1} = \frac{V_3}{V_4} \qquad (6)$$

Substituting equation (6) in equation (2),

$$\text{Work done/cycle} = mR \ln \frac{V_2}{V_1} (T_1 - T_3) \qquad (7)$$

This is net positive work done and this is always the case if the processes of a cycle proceed in a clockwise direction. External work can thus be obtained from such cycles.

If the processes proceed in an anticlockwise direction then the net work done is negative, in which case equation (7) now becomes,

$$\text{Work done/cycle} = -mR \ln \frac{V_2}{V_1} (T_1 - T_3) \qquad (8)$$

Negative work means that external work must be put in to carry out such cycles.

Now

$$\text{Thermal } \eta = \frac{\text{Heat received} - \text{Heat rejected}}{\text{Heat received}}$$

$\therefore$ from the analysis given above,

$$\text{Thermal } \eta = \frac{mRT_1 \ln (V_2/V_1) - mRT_3 \ln (V_3/V_4)}{mRT_1 \ln (V_2/V_1)}$$

$$= \frac{mR \ln (V_2/V_1)(T_1 - T_3)}{mR \ln (V_2/V_1) T_1}$$

since $V_2/V_1 = V_3/V_4$ from equation (6)

$$\therefore \text{Thermal } \eta = \frac{T_1 - T_3}{T_1} \qquad (9)$$

$$= \frac{\text{Max. abs. temp.} - \text{Min. abs. temp.}}{\text{Max. abs. temp.}} \qquad (10)$$

Now from equation (9),

$$\text{Thermal } \eta = 1 - \frac{T_3}{T_1} \qquad (11)$$

and from equations (3), (4) and (5),

$$\frac{T_1}{T_3} = \left(\frac{V_4}{V_1}\right)^{\gamma-1} = \left(\frac{V_3}{V_2}\right)^{\gamma-1} = r_v^{\gamma-1}$$

where r_v = adiabatic compression and expansion volume ratio.

∴ from equation (11),

Thermal $\eta = 1 - \dfrac{1}{r_v^{\gamma-1}}$ (12)

This thermal efficiency gives the maximum possible thermal efficiency obtainable between any two given temperature limits.

The Carnot cycle is sometimes referred to as the *constant temperature cycle* because heat is transferred during isothermal (constant temperature) processes only.

In Fig. 5.2(b) is shown the temperature–entropy diagram of the Carnot cycle which, it will be observed, appears as a rectangle.

From the T–s diagram,

Heat received from 1 to $2 = T_1(s_1 - s_2) =$ area under 1–2 (13)

Heat rejected from 3 to $4 = T_3(s_3 - s_4) =$ area under 3–4 (14)

Thermal efficiency $= \dfrac{\text{Heat received} - \text{Heat rejected}}{\text{Heat received}}$

$= \dfrac{T_1(s_2 - s_1) - T_3(s_3 - s_4)}{T_1(s_2 - s_1)}$

$= \dfrac{(T_1 - T_3)(s_2 - s_1)}{T_1(s_2 - s_1)}$, since $(s_2 - s_1) = (s_3 - s_4)$

$= \dfrac{T_1 - T_3}{T_1}$ (15)

Note that this solution for the thermal efficiency of the Carnot cycle is rather more simple than that given earlier. Also,

Work done/cycle $= \oint W =$ Heat received $-$ Heat rejected

$= (T_1 - T_2)(s_2 - s_1)$ (16)

Again note that this cycle will have the highest thermal efficiency possible within the temperature limits of the cycle. This is because it is composed entirely of reversible processes.

It is now possible to suggest that if any engine working between the same temperature limits has a thermal efficiency lower than this, then thermal efficiency improvement is theoretically possible. Most engines have a thermal efficiency much lower than the Carnot efficiency.

The ultimate aim should be an attempt to reach an efficiency as near unity (100%) as possible. Now how could this be achieved?

$$\text{Carnot thermal } \eta = \frac{\text{Max. abs. temp.} - \text{Min. abs. temp.}}{\text{Max. abs. temp.}}$$

For this equation to be a maximum it must equal unity in which case the thermal efficiency would be 100%.

Consider the minimum absolute temperature as being fixed. Then if the maximum absolute temperature is increased, so the magnitude of this equation gets larger. Eventually, as the maximum absolute temperature $\to \infty$, so the equation $\to 1$. It is quite impossible to have an infinitely high maximum absolute temperature. In any case, even with the higher temperatures which are obtainable, the materials which are at present known for use in engines will not stand up to continuous exposure to them.

Now consider the maximum absolute temperature as being fixed. If the minimum absolute temperature is reduced to zero, or, in other words, is reduced to the absolute zero of temperature, then, once again the equation equals unity, or 100% thermal efficiency. This again is impossible since all working substances will have liquefied and solidified before reaching this low temperature. In any case, the absolute zero of temperature is such a difficult temperature even to attempt to achieve that it is out of the question as the temperature of the sink of an engine.

It would appear from this that a thermal efficiency of 100% is impossible to achieve in practice. Hence all practical engines are inefficient.

However, a guide has been given as to how the thermal efficiency of engines delivering work may be improved. That is, to spread the maximum and minimum temperatures as far as possible, consistent with the satisfactory safe operation of the engine.

In section 5.1 the work ratio was defined in equation (6), which is,

$$\text{Work ratio} = \frac{\text{net work done}}{\text{positive work done}}$$

For the net positive work Carnot cycle,

$$\text{Work done/cycle} = mR \ln \frac{V_2}{V_1} (T_1 - T_3)$$

as shown in equation (7).

The positive work done during this cycle occurs during processes 1–2 and 2–3. The area under these processes shown on the P–V diagram gives the work done, thus,

$$\text{Positive work done} = mRT_1 \ln \frac{V_2}{V_1} + \frac{mR(T_1 - T_3)}{\gamma - 1} \tag{17}$$

Hence,

$$\text{Work ratio} = \frac{mR \ln \dfrac{V_2}{V_1}(T_1 - T_3)}{mRT_1 \ln \dfrac{V_2}{V_1} + \dfrac{mR(T_1 - T_3)}{\gamma - 1}}$$

$$= \frac{\ln \dfrac{V_2}{V_1}(T_1 - T_3)}{T_1 \ln \dfrac{V_2}{V_1} + \dfrac{(T_1 - T_3)}{\gamma - 1}} \tag{18}$$

In section 5.1 it was indicated that good ideal thermal efficiency together with good work ratio is required of a cycle if possible.

It is unfortunate that the Carnot cycle has a low work ratio even though it has the high ideal thermal efficiency.

Example 14

A gas operating on the Carnot cycle has the temperature limits of 300 °C and 40 °C. Determine the thermal efficiency of the cycle.

For the Carnot cycle,

$$\text{Thermal } \eta = \frac{T_1 - T_2}{T_1}$$

and, $T_1 = 300 + 273 = \underline{573 \text{ K}}$

$T_2 = 40 + 273 = \underline{313 \text{ K}}$

$\therefore$ Thermal $\eta = \dfrac{573 - 313}{573}$

$= \dfrac{260}{573}$

$= \underline{0.453\ 8}$

$= 0.453\ 8 \times 100\%$

$= \underline{45.38\%}$

Example 15

A Carnot cycle using a gas has a lower temperature limit of 30 °C and a thermal efficiency of 52%. Determine the upper temperature limit.

For a Carnot cycle,

$$\text{Thermal } \eta = \frac{T_1 - T_2}{T_1}$$

$\therefore T_1 \times$ Thermal $\eta = T_1 - T_2$

from which, $T_1 - (T_1 \times \text{thermal } \eta) = T_2$

hence, $T_1(1 - \text{thermal } \eta) = T_2$

or, $$T_1 = \frac{T_2}{(1 - \text{thermal } \eta)}$$

$T_2 = 30 + 273 = \underline{303 \text{ K}}$

$\therefore T_1 = \dfrac{303}{1 - 0.52} = \dfrac{303}{0.48} = \underline{631.25 \text{ K}}$

$t_1 = 631.25 - 273 = \underline{358.25 \,°\text{C}}$

Example 16

0.5 kg of air is at an initial pressure of 1.65 MN/m² and a temperature of 200 °C. The air is taken through a Carnot cycle in which the volume ratio for the isothermal processes is 2.8 and for the adiabatic processes the volume ratio is 2.1. The initial isothermal process is an expansion.
Determine:
(a) the pressure, volume and temperature at each corner of the cycle;
(b) the thermal efficiency of the cycle;
(c) the work done/cycle;
(d) the work ratio.
Take, $c_p = 1.005$ kJ/kgK, $c_v = 0.72$ kJ/kgK.

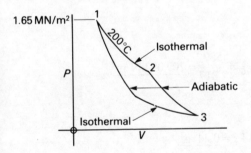

(a) For a gas,

$c_p - c_v = R = 1.005 - 0.72 = \underline{0.285 \text{ kJ/kgK}}$

$\gamma = \dfrac{c_p}{c_v} = \dfrac{1.005}{0.72} = \underline{1.396}$

$P_1 V_1 = mRT_1$

$\therefore V_1 = \dfrac{mRT_1}{P_1}$ and $T_1 = 200 + 273 = \underline{473 \text{ K}}$

$$= \frac{0.5 \times 0.285 \times 473}{1.65 \times 10^3} \quad \text{(Note pressure required in kN/m}^2\text{)}$$

$$= \underline{0.040\ 9\ \text{m}^3}$$

$\therefore\ P_1 = 1.65\ \text{MN/m}^2,\ V_1 = 0.040\ 9\ \text{m}^3,\ t_1 = 200\ °\text{C}$

$t_2 = t_1$ since process 1–2 is isothermal

$\therefore\ t_2 = \underline{200\ °\text{C}}$

$\dfrac{V_2}{V_1} = 2.8$

$\therefore\ V_2 = 2.8 V_1 = 2.8 \times 0.040\ 9 = \underline{0.114\ 5\ \text{m}^3}$

For the isothermal process 1–2,

$P_1 V_1 = P_2 V_2$

$\therefore\ P_2 = P_1 \dfrac{V_1}{V_2} = \dfrac{1.65}{2.8} = \underline{0.589\ \text{MN/m}^2} = \underline{589\ \text{kN/m}^2}$

$\therefore\ P_2 = 589\ \text{kN/m}^2,\ V_2 = 0.114\ 5\ \text{m}^3,\ t_2 = 200\ °\text{C}$

For the adiabatic process 2–3,

$$\frac{T_2}{T_3} = \left(\frac{V_3}{V_2}\right)^{\gamma-1}$$

$\therefore\ T_3 = T_2 \left(\dfrac{V_2}{V_3}\right)^{\gamma-1} = \dfrac{473}{2.1^{1.396-1}} = \dfrac{473}{2.1^{0.396}} = \dfrac{473}{1.342} = \underline{352.46\ \text{K}}$

$t_3 = \underline{79.46\ °\text{C}}$

$P_2 V_2^\gamma = P_3 V_3^\gamma$

$\therefore\ P_3 = P_2 \left(\dfrac{V_2}{V_3}\right)^\gamma = \dfrac{589}{2.1^{1.396}} = \dfrac{589}{2.817} = \underline{209.09\ \text{kN/m}^2}$

$\dfrac{V_3}{V_2} = 2.1$

$\therefore\ V_3 = 2.1 V_2 = 2.1 \times 0.114\ 5 = \underline{0.24\ \text{m}^3}$

$\therefore\ P_3 = 209.09\ \text{kN/m}^2;\ V_3 = 0.24\ \text{m}^3.\ t_3 = 79.46\ °\text{C}$

For the isothermal process 3–4,

$t_3 = t_4 = \underline{79.46\ °\text{C}}$

$\dfrac{V_3}{V_4} = 2.8$

$\therefore\ V_4 = \dfrac{V_3}{2.8} = \dfrac{0.24}{2.8} = \underline{0.086\ \text{m}^3}$

$P_3 V_3 = P_4 V_4$

$\therefore P_4 = P_3 \dfrac{V_3}{V_4} = 209.09 \times 2.8 = \underline{585.45 \text{ kN/m}^2}$

$\therefore P_4 = 585.45 \text{ kN/m}^2, V_4 = 0.086 \text{ m}^3, t_4 = 79.46 \text{ °C}.$

(b) Thermal $\eta = \dfrac{T_1 - T_3}{T_1}$

$= \dfrac{473 - 352.46}{473} = 0.255$

$= 0.255 \times 100 = \underline{25.5\%}$

Alternatively,

Thermal $\eta = 1 - \dfrac{1}{r_v^{\gamma-1}}$ and $r_v = \dfrac{V_3}{V_2} = \dfrac{0.24}{0.114\,5} = \underline{2.096}$

$= 1 - \dfrac{1}{2.096^{1.396-1}} = 1 - \dfrac{1}{2.096^{0.396}}$

$= 1 - \dfrac{1}{1.341} = 1 - 0.745$

$= \underline{0.255}$

(c) Work done/cycle $= mR \ln \dfrac{V_2}{V_1}(T_1 - T_3)$

$= 0.5 \times 0.285 \ln 2.8 (473 - 352.46)$

$= 0.5 \times 0.285 \times 1.03 \times 120.54$

$= \underline{17.7 \text{ kJ}}$

Alternatively,

Work done/cycle = Heat received × thermal η

$= mRT_1 \ln \dfrac{V_2}{V_1} \times 0.255$

$= 0.5 \times 0.285 \times 473 \ln 2.8 \times 0.255$

$= 0.5 \times 0.285 \times 473 \times 1.03 \times 0.255$

$= \underline{17.7 \text{ kJ}}$

(d) Work ratio $= \dfrac{\ln\dfrac{V_2}{V_1}(T_1 - T_3)}{T_1 \ln\dfrac{V_2}{V_1} + \dfrac{T_1 - T_3}{\gamma - 1}}$

$= \dfrac{1.03(120.54)}{(1.03 \times 473) + \dfrac{120.54}{0.396}} = \dfrac{124.156}{487.19 + 304.39}$

$= \dfrac{124.156}{791.58}$

$= \underline{0.157}$ [Note that this is a very low work ratio]

5.3 The constant volume cycle

This cycle is sometimes referred to as the Otto cycle after Dr N. A. Otto, a German scientist, who successfully applied the cycle to an internal combustion engine in 1876. The cycle was originally developed by a Frenchman named Beau de Rochas in 1862.

In Fig. 5.3(a) is represented the P–V diagram of the constant volume cycle. In Fig. 5.3(b) is represented the corresponding T–s diagram. The cycle is arranged as follows:

1–2   Adiabatic compression of the gas according to the law $PV^\gamma = C$.
   Pressure increases from P_1 to P_2.
   Volume decreases from V_1 to V_2.
   Temperature increases from T_1 to T_2.
   Entropy remains constant at $s_1 = s_2$.
2–3   Constant volume heat addition.
   Volume remains constant at $V_2 = V_3$.
   Pressure increases from P_2 to P_3.

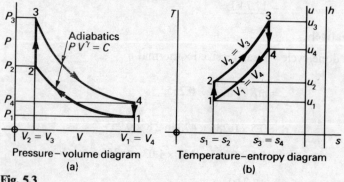

Pressure–volume diagram
(a)

Temperature–entropy diagram
(b)

Fig. 5.3

Temperature increases from T_2 to T_3.
Entropy increases from s_2 to s_3.
3–4 Adiabatic expansion of the gas according to the law $PV^\gamma = C$.
Pressure decreases from P_3 to P_4.
Volume increases from V_3 to V_4.
Temperature decreases from T_3 to T_4.
Entropy remains constant at $s_3 = s_4$.
4–1 Constant volume heat rejection.
Volume remains constant at $V_4 = V_1$.
Pressure decreases from P_4 to P_1.
Temperature decreases from T_4 to T_1.
Entropy decreases from s_4 to s_1.

This process completes the cycle and returns the gas to its original state at 1.

An analysis of the properties P, V and T at state points 1, 2, 3 and 4 will now be made and it will be assumed that P_1, V_1 and T_1 are known.

1. P_1, V_1, T_1.
2. Assume that the volume ratio V_2/V_1 is known.

In this connection it should be noted that this cycle is the one which theoretically most closely follows the processes which occur in a petrol engine. The ratio V_1/V_2 for a petrol engine is often referred to as the 'compression ratio' of the engine. Note that the ratio V_1/V_2 is, in fact, a volume ratio.

In a petrol engine, the volume V_2 is often referred to as the *clearance* volume.

$T_1/T_2 = (V_2/V_1)^{\gamma-1}$

$\therefore \quad T_2 = T_1 \left(\dfrac{V_1}{V_2}\right)^{\gamma-1} = T_1 r_v^{\gamma-1}$

where, $r_v = V_1/V_2 =$ Adiabatic compression, volume ratio,
$= V_4/V_3 =$ Adiabatic expansion, volume ratio.

Also $P_1 V_1^\gamma = P_2 V_2^\gamma$,

$\therefore \quad P_2 = P_1 \left(\dfrac{V_1}{V_2}\right)^\gamma = P_1 r_v^\gamma$

3. $V_3 = V_2$, since the volume remains constant.

$P_3/T_3 = P_2/T_2$,

$\therefore \quad T_3 = T_2 \dfrac{P_3}{P_2} = \dfrac{P_3}{P_2} T_1 r_v^{\gamma-1}$ from 2

4. $T_3/T_4 = (V_4/V_3)^{\gamma-1}$,

$$\therefore T_4 = \frac{T_3}{r_v^{\gamma-1}} = \frac{P_3}{P_2} T_1 \frac{r_v^{\gamma-1}}{r_v^{\gamma-1}} = \frac{P_3}{P_2} T_1, \quad \text{from 3}$$

Also, $P_4 V_4^\gamma = P_3 V_3^\gamma$,

$$\therefore P_4 = P_3 \left(\frac{V_3}{V_4}\right)^\gamma = \frac{P_3}{r_v^\gamma}$$

Also, from the constant volume process 4–1, $P_4/T_4 = P_1/T_1$,

$$\therefore T_4 = \frac{P_4}{P_1} T_1 = \frac{P_3}{P_2} T_1, \quad \text{from above.}$$

From this it follows that $P_4/P_1 = P_3/P_2$.

The work done during the cycle may be obtained as follows:
Process 3–4 is an expansion which gives positive work done.
Process 1–2 is a compression which gives negative work done.
The net work done is the sum of the work done by these two processes.

$$\therefore \text{Net work done} = \oint W = \text{area under 3–4} - \text{area under 1–2}$$

$$= \text{area 1 2 3 4} = \text{area of diagram}$$

$$\therefore \oint W = \frac{(P_3 V_3 - P_4 V_4)}{\gamma - 1} - \frac{(P_2 V_2 - P_1 V_1)}{\gamma - 1}$$

$$= \frac{(P_3 V_3 - P_4 V_4) - (P_2 V_2 - P_1 V_1)}{\gamma - 1} \tag{1}$$

$$= \frac{mR}{\gamma - 1} \{(T_3 - T_4) - (T_2 - T_1)\} \tag{2}$$

(since $PV = mRT$)

Alternatively, the cycle work done can be determined using the equation

$$\oint W = \oint Q$$

or

$$\text{Cycle work done} = \oint W = \text{Heat received} - \text{Heat rejected} \tag{3}$$

It should be noted that this equation (3) will hold good for any cycle.

Now in this cycle heat is received and rejected only during constant volume processes. Hence the name, the constant volume cycle.

There are also two adiabatic processes, but during an adiabatic process no heat is received or rejected. Hence the adiabatics do not appear when the discussion is on heat received or rejected. This will still be the same in any other cycle where adiabatic processes appear.

In this cycle,

Heat is received from $2\text{--}3 = mc_v(T_3 - T_2)$ (4)

Heat is rejected from $4\text{--}1 = mc_v(T_4 - T_1)$ (5)

$$\therefore \oint W = mc_v(T_3 - T_2) - mc_v(T_4 - T_1)$$

$$= mc_v\{(T_3 - T_2) - (T_4 - T_1)\} \quad (6)$$

The thermal efficiency of the cycle may be determined from the equation.

$$\text{Thermal efficiency} = \frac{\text{Heat received} - \text{Heat rejected}}{\text{Heat received}} \quad (7)$$

$$= 1 - \frac{\text{Heat rejected}}{\text{Heat received}} \quad (8)$$

$$= 1 - \frac{mc_c(T_4 - T_1)}{mc_v(T_3 - T_2)}$$

and assuming that c_v remains constant

$$\text{Thermal } \eta = 1 - \frac{T_4 - T_1}{T_3 - T_2} \quad (9)$$

This equation (9) gives the thermal efficiency in terms of temperature.

Also, substituting temperatures in terms of T_1 in equation (9)

$$\text{Thermal } \eta = 1 - \frac{(P_3/P_2)T_1 - T_1}{(P_3/P_2)T_1 r_v^{\gamma-1} - T_1 r_v^{\gamma-1}}$$

$$= 1 - \frac{T_1(P_3/P_2 - 1)}{T_1 r_v^{\gamma-1}(P_3/P_2 - 1)}$$

$$= 1 - \frac{1}{r_v^{\gamma-1}} \quad (10)$$

Now consider the adiabatics 1–2 and 3–4.

$$\frac{T_2}{T_1} = \left(\frac{V_1}{V_2}\right)^{\gamma-1} \quad \text{and} \quad \frac{T_3}{T_4} = \left(\frac{V_4}{V_3}\right)^{v-1}$$

But $V_1/V_2 = V_4/V_3 = r_v$,

$$\therefore \frac{T_2}{T_1} = \frac{T_3}{T_4} = r_v^{\gamma-1} \quad \text{or} \quad \frac{T_1}{T_2} = \frac{T_4}{T_3} = \left(\frac{1}{r_v}\right)^{v-1}$$

$\therefore$ from equation (10),

Thermal $\eta = 1 - \dfrac{T_1}{T_2}$ (11)

$ = 1 - \dfrac{T_4}{T_3}$ (12)

Note that from equation (7),

Heat received − Heat rejected = Heat received × Thermal η (13)

Substituting equation (3) into equation (13),

$$\oint W = \text{Heat received} \times \text{Thermal } \eta \tag{14}$$

This is another way in which the work done may be determined.

In the case of this cycle,

$$\oint W = mc_v(T_3 - T_2) \times \text{Thermal } \eta \tag{15}$$

The cycle can also be analysed by using the temperature–entropy chart.

By tracing the points of the cycle round the chart, the various properties P, V, T, u, h at stage points 1, 2, 3 and 4 can be found.

Now for the constant volume process, it has been shown that,

Heat transferred = Change of internal energy (see *Thermodynamics Level 3*, section 4.6]

Hence, from this,

Heat received from 2–3 = $u_3 - u_2$ (16)

Heat rejected from 4–1 = $u_4 - u_1$ (17)

From equations (3), (16) and (17), therefore,

$$\oint W = (u_3 - u_2) - (u_4 - u_1) \tag{18}$$

From equations (8), (16) and (17), therefore,

Thermal $\eta = 1 - \dfrac{(u_4 - u_1)}{(u_3 - u_2)}$ (19)

It must be remembered here, however, that since the chart is made out for unit mass of gas, then equations (16), (17) and (18) will give values/unit mass/cycle.

On a practical note concerning this cycle, note that for thermal efficiency increase so $r_v = V_1/V_2$ must increase, (i.e. the 'compression ratio' must increase – see earlier note and equation 10). In a practical engine the ability to increase r_v is limited because of high material loading, high temperatures and fuel combustion problems. In the case of the petrol engine the addition of tetraethyl lead to the fuel, for example, enabled the fuel to be used at higher compression ratios, (it assisted in helping to prevent 'knocking' or 'pinking'). The use of tetraethyl lead is now in disfavour because of atmospheric pollution and possible damage to health.

Example 16
An engine working on a constant volume cycle has a bore of 100 mm and a stroke of 120 mm. The clearance volume of the engine is 0.12 litre. If the relative efficiency of the engine = 55%, determine the actual thermal efficiency of engine. Take $\gamma = 1.4$

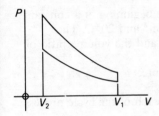

For the engine, stroke volume $= V_1 - V_2$

$$= \frac{\pi d^2}{4} \times l$$

where $d = $ bore, $l = $ stroke

$$= \frac{\pi \times 100^2}{4} \times 120$$

$$= 7\,854 \times 120$$

$$= 942\,480 \text{ mm}^3$$

Clearance volume $= V_2 = 0.12 \times 10^6 = \underline{120\,000 \text{ mm}^3}$

Total volume $= V_1 = (V_1 - V_2) + V_2 = 942\,480 + 120\,000$

$$= \underline{1\,062\,480 \text{ mm}^3}$$

Ideal thermal $\eta = 1 - \dfrac{1}{r_v^{\gamma-1}}$

and $r_v = \dfrac{V_1}{V_2} = \dfrac{1\,062\,480}{120\,000} = 8.85$

∴ Ideal thermal $\eta = 1 - \dfrac{1}{8.85^{1.4-1}} = 1 - \dfrac{1}{8.85^{0.4}}$

$$= 1 - \frac{1}{2.342} = 1 - 0.418$$

$$= \underline{0.582}$$

$$\text{Relative } \eta = \frac{\text{actual thermal } \eta}{\text{ideal thermal } \eta}$$

∴ actual thermal η = Relative thermal η × ideal thermal η

$$= 0.55 \times 0.582$$

$$= \underline{0.32} = 0.32 \times 100\% = \underline{32\%}$$

Example 17

The pressure, volume and temperature at the beginning of a constant volume cycle using air are 100 kN/m^2, 0.002 m^2 and $20\,°C$. The maximum temperature of the cycle is $800\,°C$ and the volume ratio of the cycle is 8:1.

The cycle is repeated 2 500 times per minute.

Determine:
(a) the pressure, volume and temperature at each of the cycle process change points;
(b) the work done/minute;
(c) the thermal efficiency of the cycle;
(d) the mean effective pressure of the cycle.

Take, $c_p = 1.006 \text{ kJ/kgK}$; $c_v = 0.717 \text{ kJ/kgK}$.

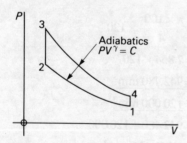

(a) $P_1 = 100 \text{ kN/m}^2$, $V_1 = 0.002 \text{ m}^3$, $t_1 = 20\,°C$, all given.

For process 1–2.

$$P_1 V_1^\gamma = P_2 V_2^\gamma \quad \text{and} \quad \gamma = c_p/c_v = 1.006/0.717 = \underline{1.403}.$$

$$\therefore P_2 = P_1 \left(\frac{V_1}{V_2}\right)^\gamma \quad \text{and} \quad V_1/V_2 = 8$$

$$= 100 \times 8^{1.403} = 100 \times 18.494$$

$$= \underline{1\,849.4\,\text{kN/m}^2}$$

$$\frac{T_2}{T_1} = \left(\frac{V_1}{V_2}\right)^{\gamma-1} \quad \text{and} \quad T_1 = 20 + 273 = \underline{293\,\text{K}}$$

$$\therefore\ T_2 = T_1\left(\frac{V_1}{V_2}\right)^{\gamma-1}$$

$$= 293 \times 8^{1.403-1}$$

$$= 293 \times 8^{0.403} = 293 \times 2.312$$

$$= \underline{677.42\,\text{K}}$$

$$t_2 = 677.42 - 273$$

$$= \underline{404.42\,°\text{C}}$$

$$V_2 = 0.002/8$$

$$= \underline{0.000\,25\,\text{m}^3}$$

For process 2–3

$$V_3 = V_2 = \underline{0.000\,25\,\text{m}^3}$$

$$\frac{P_3}{T_3} = \frac{P_2}{T_2} \quad \text{and} \quad T_3 = 800 + 273 = \underline{1\,073\,\text{K}}$$

$$\therefore\ P_3 = P_2 \frac{T_3}{T_2}$$

$$= 1\,849.4 \times \frac{1\,073}{677.42}$$

$$= \underline{2\,929.36\,\text{kN/m}^2}$$

For process 3–4

$$P_3 V_3^\gamma = P_4 V_4^\gamma$$

$$\therefore\ P_4 = P_3\left(\frac{V_3}{V_4}\right)^\gamma = \frac{2\,929.36}{8^{1.403}} = \frac{2\,929.36}{18.494}$$

$$= \underline{158.4\,\text{kN/m}^2}$$

$$\frac{T_3}{T_4} = \left(\frac{V_4}{V_3}\right)^{\gamma-1}$$

$$\therefore\ T_4 = T_3\left(\frac{V_3}{V_4}\right)^{\gamma-1} = \frac{1\,073}{8^{0.403}} = \frac{1\,073}{2.312}$$

$$= \underline{464.1\,\text{K}}$$

$$t_3 = 464.1 - 273$$

$$= \underline{191.1\,°\text{C}}$$

$$V_4 = V_1 = \underline{0.002\,\text{m}^3}$$

(b) $\oint W$ = Heat received − Heat rejected

$$= mc_v\{(T_3 - T_2) - (T_4 - T_1)\}$$

and $m = \dfrac{P_1 V_1}{RT_1} = \dfrac{100 \times 0.002}{(1.006 - 0.717) \times 293}$

$$= \dfrac{100 \times 0.002}{0.289 \times 293} = \underline{0.002\,4\ \text{kg}}$$

$\therefore \oint W = 0.002\,4 \times 0.717\{(1\,073 - 677.42) - (464.1 - 293)\}$

$$= 0.002\,4 \times 0.717\{395.58 - 171.1\}$$

$$= 0.002\,4 \times 0.717 \times 224.48$$

$$= \underline{0.386\ \text{kJ}}$$

$\therefore$ Work done/minute $= 0.386 \times 2\,500$

$$= \underline{965\ \text{kJ}}$$

(c) Thermal $\eta = 1 - \dfrac{1}{r_v^{\gamma-1}}$

$$= 1 - \dfrac{1}{8^{0.043}} = 1 - \dfrac{1}{2.312} = 1 - 0.433$$

$$= \underline{0.567}$$

(d) Mean effective pressure $= \dfrac{\text{work done/cycle}}{V_1 - V_2}$

$$= \dfrac{0.386}{0.002 - 0.000\,25}$$

$$= \dfrac{0.386}{0.01\,75}$$

$$= \underline{220.57\ \text{kN/m}^2}.$$

5.4 The constant pressure cycle

This cycle is sometimes referred to as the Joule cycle after James Prescott Joule (1819–89), an English physicist. The unit of energy, the *Joule*, has been named after this physicist.

Joule conceived this cycle for use in an air engine.

The cycle is sometimes referred to as the Brayton cycle, so-named after George Brayton, an American engineer, who in about 1870, attempted the use of this cycle in a gas engine. It is possible, however, that John Ericsson (1803–89), developed this cycle at an earlier date than both Joule or Brayton. Of interest, John Ericsson was a Swedish engineer who spent some time in England and constructed a steam locomotive at the same time as George Stephenson. Ericsson eventually left England to live in America, where his interests centred on hot air engines.

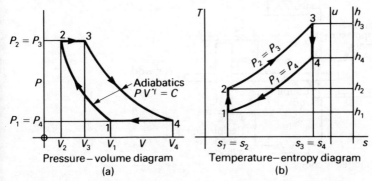

Fig. 5.4

In Fig. 5.4(a) is represented the P–V diagram of the constant pressure cycle. In Fig. 5.4(b) is represented the corresponding T–s diagram. The cycle is arranged as follows:

1–2 Adiabatic compression according to the law $PV^\gamma = C$.
   Pressure increases from P_1 to P_2.
   Temperature increases from T_1 to T_2.
   Volume decreases from V_1 to V_2.
   Entropy remains constant at $s_1 = s_2$.
2–3 Constant pressure heat addition.
   Pressure remains constant at $P_2 = P_3$.
   Temperature increases from T_2 to T_3.
   Volume increases from V_2 to V_3.
   Entropy increases from s_2 to s_3.
3–4 Adiabatic expansion according to the law $PV^\gamma = C$.
   Pressure decreases from P_3 to P_4.
   Temperature decreases from T_3 to T_4.
   Volume increases from V_3 to V_4.
   Entropy remains constant at $s_3 = s_4$.
4–1 Constant pressure heat rejection.
   Pressure remains constant at $P_4 = P_1$.
   Temperature decreases from T_4 to T_1.
   Volume decreases from V_4 to V_1.
   Entropy decreases from s_4 to s_1.
   This process completes the cycle and returns the gas to its original state.

It should be noted that in practice this cycle has been mostly considered for use in engines which do not carry out all processes in a single unit, a cylinder, for example. Compression and expansion have been arranged for in separate units. For example, Joule's concept of an air engine is illustrated diagrammatically in Fig. 5.5(a).

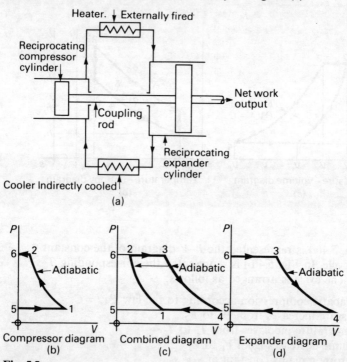

Fig. 5.5

An arrangement for the engine was that of a reciprocating air compressor connected in tandem, through a coupling rod, to a reciprocating expander. The net work output from the expander appears at the piston rod. Air from the compressor was fed through an externally fired heater in which it was expanded, theoretically at constant pressure. From the heater the air was fed to the expander. From the expander the air entered a cooler in which it was cooled, theoretically at constant pressure. The air was then fed to the compressor for recirculation. Note that the system is closed, the same air being circulated through the engine (neglecting losses). At (b) and (d) in Fig. 5.4 are shown the compressor and expander $P-V$ diagrams. The combined diagrams are shown at (c). The net diagram for the engine is given by 1 2 3 4, which, it will be recognised, is the same as that shown in Fig. 5.4(a). The net diagram 1 2 3 4 is called the *constant pressure cycle* and its area will give the theoretical net work output. Note that the expander must provide the internal work for the compressor as well as the net work output. Of further interest, it should be noted that by replacing the reciprocating compressor and expander by a rotary compressor and a turbine this same arrangement becomes the basic design for a constant pressure gas turbine.

The gas turbine will be discussed later in this section, (also see section 4.13). A further point of interest concerning the constant pressure cycle is that by reversing the airflow and by introducing work input to drive the plant, this cycle arrangement was used in the past as a refrigerator. A notable refrigerator of this type was the Bell–Coleman refrigerator of about 1880.

Referring to Fig. 5.5 an analysis of the properties at state points 1, 2, 3 and 4 can now be made. It will be assumed that P_1, V_1 and T_1 are known.

1. P_1, V_1, T_1.
2. Assume that the volume ratio V_1/V_2 is known.

$$T_1/T_2 = (V_2/V_1)^{\gamma-1},$$

$$\therefore \quad T_2 = T_1 \left(\frac{V_1}{V_2}\right)^{\gamma-1} = T_1 r_v^{\gamma-1}$$

where $r_v = V_1/V_2 =$ Adiabatic compression, volume ratio.

Also, $P_1 V_1^\gamma = P_2 V_2^\gamma$,

$$\therefore \quad P_2 = P_1 \left(\frac{V_1}{V_2}\right)^\gamma = P_1 r_v^\gamma$$

3. $P_3 = P_2$, since the pressure remains constant.

$V_3/T_3 = V_2/T_2,$

$$\therefore \quad T_3 = T_2 \frac{V_3}{V_2} = \frac{V_3}{V_2} T_1 r_v^{\gamma-1}$$

4. $T_4/T_3 = (V_3/V_4)^{\gamma-1}$,

$$\therefore \quad T_4 = T_3 \left(\frac{V_3}{V_4}\right)^{\gamma-1}$$

Now consider the adiabatics 1–2 and 3–4. Both have the same pressure ratio $P_2/P_1 = P_3/P_4$.

For adiabatic 1–2,

$$\frac{P_2}{P_1} = \left(\frac{V_1}{V_2}\right)^\gamma$$

For adiabatic 3–4,

$$\frac{P_3}{P_4} = \left(\frac{V_4}{V_3}\right)^\gamma$$

But $P_2/P_1 = P_3/P_4$. Therefore it follows that,

$$\frac{V_1}{V_2} = \frac{V_4}{V_3} = r_v = \text{Adiabatic compression and expansion, volume ratio}$$

Hence, from above,

$$T_4 = \frac{T_3}{r^{\gamma-1}} = \frac{V_3}{V_2} T_1 \frac{r_v^{\gamma-1}}{r_v^{\gamma-1}} = \frac{V_3}{V_2} T_1$$

Note also that for the constant pressure process 4–1, $V_4/T_4 = V_1/T_1$,

$$\therefore T_4 = \frac{V_4}{V_1} T_1$$

$\therefore$ it follows that

$$\frac{V_4}{V_1} = \frac{V_3}{V_2} = \text{Constant pressure processes volume ratios}$$

This could also have been obtained from the fact that since $V_1/V_2 = V_4/V_3 = r_v$, then, $V_4/V_1 = V_3/V_2$.

Also, $P_4 V_4^\gamma = P_3 V_3^\gamma$,

$$\therefore P_4 = P_3 \left(\frac{V_3}{V_4}\right)^\gamma = \frac{P_3}{r_v^\gamma}$$

The work done during the cycle may be obtained as follows:
Processes 2–3 and 3–4 are expansions and give positive work done.
Processes 4–1 and 1–2 are compressions and give negative work done.
The net work done during the cycle will be the sum of the work done during these processes.

Hence,

$$\oint W = \text{Area under 2–3} + \text{area under 3–4}$$

$$\quad - \text{area under 4–1} - \text{area under 1–2}$$

$$= P_2(V_3 - V_2) + \frac{(P_3 V_3 - P_4 V_4)}{\gamma - 1} - P_1(V_4 - V_1) - \frac{(P_2 V_2 - P_1 V_1)}{\gamma - 1}$$

$$= P_2(V_3 - V_2) - P_1(V_4 - V_1) + \frac{(P_3 V_3 - P_4 V_4) - (P_2 V_2 - P_1 V_1)}{\gamma - 1} \quad (1)$$

$$= mR(T_3 - T_2) - mR(T_4 - T_1) + \frac{mR}{\gamma - 1}\{(T_3 - T_4) - (T_2 - T_1)\}$$

$$= mR\left[(T_3 - T_2) - (T_4 - T_1) + \frac{\{(T_3 - T_4) - (T_2 - T_1)\}}{\gamma - 1}\right]$$

$$= mR\left[(T_3 - T_4) - (T_2 - T_1) + \frac{\{(T_3 - T_4) - (T_2 - T_1)\}}{\gamma - 1}\right]$$

$$= mR\{(T_3 - T_4) - (T_2 - T_1)\}\left(1 + \frac{1}{\gamma - 1}\right)$$

$$= mR\{(T_3 - T_4) - (T_2 - T_1)\}\left(\frac{\gamma - 1 + 1}{\gamma - 1}\right)$$

$$= mR \frac{\gamma}{\gamma - 1}\{(T_3 - T_4) - (T_2 - T_1)\} \tag{2}$$

Alternatively, the work done may be obtained from the equation,

Work done = Heat received − Heat rejected

In this cycle, heat is received during the constant pressure process 2–3 and rejected during constant pressure process 4–1.

No heat is received or rejected during the adiabatic processes.
Hence,

$$\oint W = mc_p(T_3 - T_2) - mc_p(T_4 - T_1)$$

$$= mc_p\{(T_3 - T_2) - (T_4 - T_1)\} \tag{3}$$

Again, the work done may be obtained from the equation,

$$\oint W = \text{Heat received} \times \text{Thermal } \eta$$

$$\therefore \quad \oint W = mc_p(T_3 - T_2) \times \text{Thermal } \eta \tag{4}$$

The thermal efficiency may be obtained as follows

$$\text{Thermal } \eta = 1 - \frac{\text{Heat rejected}}{\text{Heat received}}$$

$$= 1 - \frac{mc_p(T_4 - T_1)}{mc_p(T_3 - T_2)}$$

and assuming that c_p remains constant,

$$\text{Thermal } \eta = 1 - \frac{(T_4 - T_1)}{(T_3 - T_2)} \tag{5}$$

This equation (5) gives the thermal efficiency in terms of temperature.

Also, substituting temperatures in terms of T_1 in equation (5),

$$\text{Thermal } \eta = 1 - \frac{(V_3/V_2)T_1 - T_1}{(V_3/V_2)T_1 r_v^{\gamma-1} - T_1 r_v^{\gamma-1}}$$

$$= 1 - \frac{T_1(V_3/V_2 - 1)}{T_1 r_v^{\gamma-1}(V_3/V_2 - 1)}$$

$$= 1 - \frac{1}{r_v^{\gamma-1}} \tag{6}$$

Now consider the adiabatics 1–2 and 3–4. Each have the same pressure ratio.

$$\frac{T_2}{T_1} = \left(\frac{P_2}{P_1}\right)^{(\gamma-1)/\gamma} = \left(\frac{V_1}{V_2}\right)^{\gamma-1} = r_v^{\gamma-1}$$

also,

$$\frac{T_3}{T_4} = \left(\frac{P_3}{P_4}\right)^{(\gamma-1)/\gamma} = \left(\frac{V_4}{V_3}\right)^{\gamma-1} = r_v^{\gamma-1}$$

and since $P_2/P_1 = P_3/P_4$, then

$$\frac{T_2}{T_1} = \frac{T_3}{T_4} = r_v^{\gamma-1} \quad \text{or} \quad \frac{T_1}{T_2} = \frac{T_4}{T_3} = \frac{1}{r_v^{\gamma-1}}$$

Hence, from equation (6),

$$\text{Thermal } \eta = 1 - \frac{T_1}{T_2} \tag{7}$$

$$= 1 - \frac{T_4}{T_3} \tag{8}$$

Also, since $\dfrac{P_2}{P_1} = \dfrac{P_3}{P_4} = r_p = \text{pressure ratio}$

and $\dfrac{T_2}{T_1} = r_p^{(\gamma-1)/\gamma}$

$\dfrac{T_3}{T_4} = r_p^{(\gamma-1)/\gamma}$

∴ from equations (7) and (8)

$$\text{Thermal } \eta = 1 - \frac{1}{r_p^{(\gamma-1)/\gamma}} \tag{9}$$

Now for a constant pressure process,

Heat received or rejected = Change of enthalpy

∴ using a T–s chart,

Heat received $= h_3 - h_2$

Heat rejected $= h_4 - h_1$

$$\text{Thermal } \eta = 1 - \frac{\text{Heat rejected}}{\text{Heat received}}$$

$$= 1 - \frac{(h_4 - h_1)}{(h_3 - h_2)} \tag{10}$$

Also,

$$\oint W = \text{Heat received} - \text{Heat rejected}$$

$$= (h_3 - h_2) - (h_4 - h_1) \tag{11}$$

Once again, this will give the work done/unit mass/cycle since the chart will be made out for unit mass of gas.

As stated earlier, the constant pressure cycle is the ideal cycle used in the constant pressure gas turbine. The basic design of the constant pressure gas turbine is obtained by replacing the reciprocating compressor and expander as shown in Fig. 5.5 with a rotary compressor and turbine.

This is shown schematically in Fig. 5.6. The cycle for this arrangement is as shown on the $P-V$ and $T-s$ diagrams in Fig. 5.4.

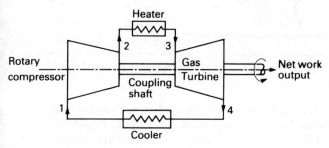

Fig. 5.6

Note that the gas turbine must provide the necessary work to run the rotary compressor. Thus,

$$\begin{matrix}\text{Net work} \\ \text{output}\end{matrix} = \begin{matrix}\text{Work done by} \\ \text{gas turbine}\end{matrix} - \begin{matrix}\text{work required by} \\ \text{rotary compressor}\end{matrix} \tag{12}$$

The arrangement shown in Fig. 5.6 has a closed circuit.
Most gas turbines, in practice, are open-circuit.
The open-circuit arrangement is illustrated in Fig. 5.7(a).

Air is taken into a rotary compressor from the atmosphere. After compression it is passed to a combustion chamber (or chambers) into which fuel (fuel oil, kerosene [paraffin] or gas) is injected and burnt continuously while the pressure remains constant.

From the combustion chamber the air (plus products of combustion) passes into a gas turbine in which it is expanded to be exhausted into the atmosphere. If the arrangement is required to produce shaft power then the turbine is designed to produce as much shaft power as possible, (see equation 12). If, however, the arrangement is required to produce jet power (as in an aircraft) then the turbine is

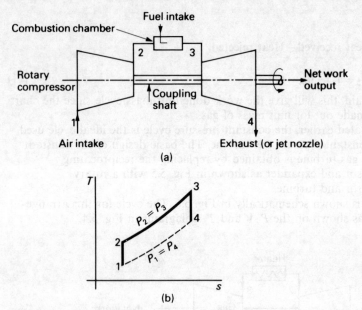

Fig. 5.7

designed to produce only the power required to run the compressor and such auxiliaries (oil pump, fuel pump, etc.) as are required.

Figure 5.7(b) illustrates the T–s diagram for the basic design of the open-circuit gas turbine.

Now, in section 4.13 it was shown that the work done in a turbine, or the work required by a rotary compressor, is equal to the change in enthalpy.

For a gas, change in specific enthalpy,

$$h_2 - h_1 = c_p(T_2 - T_1) \tag{13}$$

Hence, from equations (12) and (13), for the gas turbine,

$$\oint W = mc_p(T_3 - T_4) - mc_p(T_2 - T_1)$$

$$= mc_p\{(T_3 - T_4) - (T_2 - T_1)\} \tag{14}$$

Note that this equation could have been obtained by rearranging equation (3).

Note, also, that the compression and expansion in Figs 5.4(b) and 5.7(b) have been shown as being frictionless.

If the effect of friction is included then the T–s diagram will become as illustrated in Fig. 5.8.

In this case the isentropic efficiency of both the compressor and the turbine will be required, (see section 4.13).

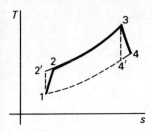

Fig. 5.8

It should be noted that since, in the arrangement as described, the cycle is open-circuit and the fuel is burnt continuously at constant pressure, then the description is given as an open-circuit, continuous combustion, constant pressure gas turbine.

Another point arises with the constant pressure cycle is that of the maximum possible work output within given temperature limits.

Consider Fig. 5.9.

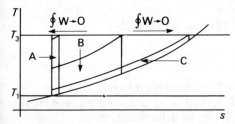

Fig. 5.9

Here are illustrated three cycles A, B and C. The temperature limits of each cycle are low temperature T_1 and high temperature T_3. The maximum pressure ratio possible between these temperature limits is given by the equation,

$$\frac{P_3}{P_1} = \left(\frac{T_3}{T_1}\right)^{\gamma/(\gamma-1)} = r_{p\max} \tag{15}$$

Consider cycle A shown in Fig. 5.9. Here the pressure ratio $r_p \rightarrow r_{p\max}$ but the net cycle work $\oint W \rightarrow 0$.

Consider cycle C. Here the pressure ratio $r_p \rightarrow 0$ and, once again, the net cycle work $\oint W \rightarrow 0$.

Somewhere in between, therefore, a cycle must exist in which $\oint W =$ a maximum. This is illustrated as cycle B. The problem now is to determine the pressure ratio, r_p, which will give the maximum net cycle work.

Now equation (14) gives the net cycle work as,

$$\oint W = mc_p\{(T_3 - T_4) - (T_2 - T_1)\}$$

The pressure ratio of the cycle is given by,

$$r_p = \frac{P_2}{P_1} = \frac{P_3}{P_4}$$

and,

$$\frac{T_2}{T_1} = \left(\frac{P_2}{P_1}\right)^{(\gamma-1)/\gamma} = r_p^{(\gamma-1)/\gamma}$$

$$\therefore T_2 = T_1 r_p^{(\gamma-1)/\gamma} \tag{17}$$

also,

$$\frac{T_3}{T_4} = \left(\frac{P_3}{P_4}\right)^{(\gamma-1)/\gamma} = r_p^{(\gamma-1)/\gamma}$$

$$\therefore T_4 = \frac{T_3}{r_p^{(\gamma-1)/\gamma}} \tag{18}$$

Substituting equations (17) and (18) into equation (14),

$$\oint W = mc_p\left\{\left(T_3 - \frac{T_3}{r_p^{(\gamma-1)/\gamma}}\right) - (T_1 r_p^{(\gamma-1)/\gamma} - T_1)\right\} \tag{19}$$

Now, assuming m, c_p, T_1 and T_3 to be constant and by differentiating equation (19) with respect to r_p and equating to zero, the value of r_p to give maximum $\oint W$ can be determined, thus,

$$\frac{d}{dr_p}\left[mc_p\left\{\left(T_3 - \frac{T_3}{r_p^{(\gamma-1)/\gamma}}\right) - (T_1 r_p^{(\gamma-1)/\gamma})\right\}\right] = 0 \quad \text{for max. } \oint W$$

or,

$$-\left\{-\left(\frac{\gamma-1}{\gamma}\right)\right\}T_3 r_p^{-(\gamma-1/\gamma)-1} - \left(\frac{\gamma-1}{\gamma}\right)T_1 r_p^{(\gamma-1/\gamma)-1} = 0$$

from which,

$$\left(\frac{\gamma-1}{\gamma}\right)T_3 r_p^{(-2\gamma+1)/\gamma} - \left(\frac{\gamma-1}{\gamma}\right)T_1 r_p^{-1/\gamma} = 0$$

$$\left(\frac{\gamma-1}{\gamma}\right)T_3 r_p^{(-2\gamma+1)/\gamma} = \left(\frac{\gamma-1}{\gamma}\right)T_1 r_p^{-1/\gamma}$$

or,

$$T_3 r_p^{(-2\gamma+1)/\gamma} = T_1 r_p^{-1/\gamma}$$

hence,
$$\frac{r_p^{-1/\gamma}}{r_p^{(-2\gamma+1)/\gamma}} = \frac{T_3}{T_1}$$

from which,
$$r_p^{-(1/\gamma)-[-(2\gamma+1)/\gamma]} = \frac{T_3}{T_1}$$

or,
$$r_p^{(2\gamma-2)/\gamma} = \frac{T_3}{T_1}$$

$$r_p^{2[(\gamma-1)/\gamma]} = \frac{T_3}{T_1}$$

hence,
$$r_p = \left(\frac{T_3}{T_1}\right)^{\gamma/[2(\gamma-1)]}$$

or,
$$r_p = \sqrt{\left(\frac{T_3}{T_1}\right)^{\gamma(\gamma-1)}} \qquad (21)$$

Substituting equation (15) into equation (21) shows that within the given temperature limits of T_1 and T_3 the maximum net cycle work is obtained when the pressure ratio of the cycle is given by the expression

$$r_p = \sqrt{r_{p\max}} \qquad (22)$$

Note that from equations (15) and (17) and (22)

$$T_2 = T_1 r_p^{(\gamma-1)/\gamma} = \frac{T_3}{r_{p\max}^{(\gamma-1)/\gamma}} r_p^{(\gamma-1)/\gamma} = \frac{T_3}{r_p^{[2(\gamma-1)/\gamma]}} r_p^{(\gamma-1)/\gamma} = \frac{T_3}{r_p^{(\gamma-1)/\gamma}}$$

But from equation (18)

$$T_4 = \frac{T_3}{r_p^{(\gamma-1)/\gamma}}$$

Hence it follows that when the maximum net cycle work is obtained within the given temperature limits,

$$T_2 = T_4 \qquad (23)$$

Note further that for this cycle,

$$\text{Thermal } \eta = \frac{(T_3 - T_4) - (T_2 - T_1)}{(T_3 - T_2)}$$

and,

$$\text{Work ratio} = \frac{(T_3 - T_4) - (T_2 - T_1)}{(T_3 - T_4)}$$

Hence, since $T_2 = T_4$ for the pressure ratio which gives maximum net work output then, for this case, it follows that,

Thermal $\eta =$ Work ratio   (24)

Example 18

In an ideal constant pressure cycle, using air, the overall volume ratio is 8.5:1. The pressure, volume and temperature of the air at the beginning of the adiabatic compression are 120 kN/m^2, 0.02 m^3 and $20\,°C$, respectively. At the end of the adiabatic compression the pressure is 1.1 MN/m^2. For the air, take, $\gamma = 1.4$, $c_p = 1.006 \text{ kJ/kgK}$. Determine:
(a) the pressure, volume and temperature at the cycle change points;
(b) the work done/cycle;
(c) the thermal efficiency of the cycle;
(d) the work ratio of the cycle;
(e) the mean effective pressure of the cycle.

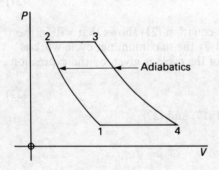

(a) $P_1 = 120 \text{ kN/m}^2$, $V_1 = 0.02 \text{ m}^3$, $t_1 = 20\,°C$, given.

$P_1 V_1^\gamma = P_2 V_2^\gamma$

$7 \quad V_2 = V_1 \left(\dfrac{P_1}{P_2}\right)^{1/\gamma} = 0.02 \times \left(\dfrac{120}{1\,100}\right)^{1/1.4}$

$\quad = \left(\dfrac{0.02}{9.17}\right)^{1/1.4} = \dfrac{0.02}{4.87}$

$V_2 = \underline{0.004\,1 \text{ m}^3}$

$$\frac{T_2}{T_1} = \left(\frac{P_2}{P_1}\right)^{(\gamma-1)/\gamma}$$

$$\therefore\ T_2 = T_1\left(\frac{P_2}{P_1}\right)^{(\gamma-1)/\gamma} \quad \text{and} \quad T_1 = 20 + 273 = \underline{293\text{ K}}$$

$$= 293 \times \left(\frac{1\,100}{120}\right)^{(1.4-1)/1.4} = 293 \times 9.17^{1/3.5}$$

$$= 293 \times 1.88 = \underline{550.8\text{ K}}$$

$$t_2 = 550.8 - 273 = \underline{277.8\,°\text{C}}$$

$$\frac{V_4}{V_2} = 8.5$$

$$\therefore\ V_4 = 8.5 V_2 = 8.5 \times 0.004\,1 = \underline{0.035\text{ m}^3}$$

$$P_4 = P_1 = \underline{120\text{ kN/m}^2}$$

$$\frac{V_4}{T_4} = \frac{V_1}{T_1}, \quad \text{since } P_1 = P_4$$

$$\therefore\ T_4 = T_1\frac{V_4}{V_1} = 293 \times \frac{0.035}{0.02} = \underline{512.75\text{ K}}$$

$$t_4 = 512.75 - 273 = \underline{239.75}$$

$$P_3 = P_2 = 1.1\ \underline{\text{MN/m}^2}$$

$$P_3 V_3^\gamma = P_4 V_4^\gamma$$

$$\therefore\ V_3 = V_4\left(\frac{P_4}{P_3}\right)^{1/\gamma} = 0.035 \times \left(\frac{120}{1\,100}\right)^{1/1.4} = \frac{0.035}{4.87}$$

$$= \underline{0.007\,2\text{ m}^3}$$

$$\frac{V_3}{T_3} = \frac{V_2}{T_2}, \quad \text{since } P_2 = P_3$$

$$\therefore\ T_3 = T_2\frac{V_3}{V_2} = 550.8 \times \frac{0.007\,2}{0.004\,1} = \underline{967.3\text{ K}}$$

$$t_3 = 967.3 - 273 = \underline{694.3\,°\text{C}}$$

Note that $\dfrac{V_4}{V_1} = \dfrac{V_3}{V_2} = \underline{1.75}$

Hence,

$P_1 = 120\text{ kN/m}^2 \qquad V_1 = 0.02\text{ m}^3 \qquad t_1 = 20\,°\text{C}$
$P_2 = 1.1\text{ MN/m}^2 \qquad V_2 = 0.004\,1\text{ m}^3 \qquad t_2 = 277.8\,°\text{C}$
$P_3 = 1.1\text{ MN/m}^2 \qquad V_3 = 0.007\,2\text{ m}^3 \qquad t_3 = 694.3\,°\text{C}$
$P_4 = 120\text{ kN/m}^2 \qquad V_4 = 0.035\text{ m}^3 \qquad \mathbf{t_4} = 239.75\,°\text{C}$

(b) $c_p - c_v = R$ and $c_p/c_v = \gamma$ from which $c_v = c_p/\gamma$

$$\therefore c_p - \frac{c_p}{\gamma} = R$$

$$c_p\left(1 - \frac{1}{\gamma}\right) = R$$

$$c_p\left(\frac{\gamma - 1}{\gamma}\right) = R$$

$$\therefore R = 1.006\left(\frac{1.4 - 1}{1.4}\right) = \frac{1.006}{3.5} = \underline{0.287 \text{ kJ/kgK}}$$

$P_1 V_1 = mRT_1$

$$\therefore m = \frac{P_1 V_1}{RT_1} = \frac{120 \times 0.02}{0.287 \times 293} = \underline{0.028\ 5 \text{ kg}}$$

$$\oint W = mc_p\{(T_3 - T_2) - (T_4 - T_1)\}, \quad \text{from equation (3)}$$

$$= 0.028\ 5 \times 1.006\{(967.3 - 550.8) - (512.75 - 293)\}$$

$$= 0.028\ 5 \times 1.006\{416.5 - 219.75\}$$

$$= 0.028\ 5 \times 1.006 \times 196.75$$

$$= \underline{5.64 \text{ kJ}}$$

(c) Thermal $\eta = 1 - \dfrac{(T_4 - T_1)}{(T_3 - T_2)}$, from equation (5)

$$= 1 - \frac{(512.75 - 293)}{(967.3 - 550.8)}$$

$$= 1 - \frac{219.75}{416.5} = 1 - 0.528$$

$$= 0.472 = 0.472 \times 100\% = \underline{47.2\%}$$

(d) Referring to the P–V diagram,

Positive work of the cycle

= area under 1–2 + area under 3–4

$$= P_2(V_3 - V_2) + \frac{(P_3 V_3 - P_4 V_4)}{\gamma - 1}$$

$$= 1\ 100(0.007\ 2 - 0.004\ 1) + \frac{\{(1\ 100 \times 0.007\ 2) - (120 \times 0.035)\}}{1.4 - 1}$$

$$= (1\ 100 \times 0.003\ 1) + \frac{(7.92 - 4.2)}{0.4}$$

$= 3.41 + 9.3 = \underline{12.71 \text{ kJ}}$

Work ratio $= \dfrac{\text{work done/cycle}}{\text{positive work of the cycle}}$

$= \dfrac{5.64}{12.71}$

$= \underline{0.444}$

(e) Mean effective pressure $= \dfrac{\text{work done/cycle}}{V_4 - V_2}$

$= \dfrac{5.64}{0.035 - 0.004\,1} = \dfrac{5.64}{0.030\,9}$

$= \underline{182.5 \text{ kN/m}^2}$

Example 19
An open circuit, continuous combustion, constant pressure, gas turbine receives air at a pressure of 95 kN/m² and a temperature of 12 °C. The air is compressed adiabatically in a rotary compressor to a pressure of 570 kN/m² with an isentropic efficiency of 82%. The compressed air passes through a combustion chamber at constant pressure in which its temperature is raised to 850 °C. The air then passes through a gas turbine in which it is expanded adiabatically to an exhaust pressure of 95 kN/m² with an isentropic efficiency of 84%.

The air flows through the plant at a rate of 3 kg/s. Neglect of the mass of fuel and take $\gamma = 1.4$ and $c_p = 1.006$ kJ/kgK, throughout. Determine:
(a) the net power output of the turbine assuming that the turbine is coupled to run the compressor;
(b) the thermal efficiency;
(c) the work ratio.

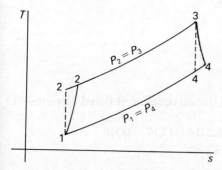

(a) $\dfrac{T_{2'}}{T_1} = \left(\dfrac{P_2}{P_1}\right)^{(\gamma-1)/\gamma}$ and $T_1 = 12 + 273 = \underline{285 \text{ K}}$

$\therefore\ T_{2'} = T_1\left(\dfrac{P_2}{P_1}\right)^{(\gamma-1)/\gamma} = 285 \times \left(\dfrac{570}{95}\right)^{(1.4-1)/1.4} = 285 \times 6^{0.4/1.4}$

$= 285 \times 6^{1/3.5} = 285 \times 1.668$

$= \underline{475.38 \text{ K}}$

For the compressor,

Isentropic $\eta_c = \dfrac{T_{2'} - T_1}{T_2 - T_1}$, (see section 4.13)

$\therefore\ T_2 - T_1 = \dfrac{T_{2'} - T_1}{\eta_c}$

$= \dfrac{475.38 - 285}{0.82} = \dfrac{190.38}{0.82} = \underline{232.17 \text{ K}}$

from which,

$T_2 = 273.17 + 285 = \underline{517.17 \text{ K}}$

For the turbine,

Isentropic $\eta_t = \dfrac{T_3 - T_4}{T_3 - T_{4'}}$, (see section 4.13)

and, $\dfrac{T_{4'}}{T_3} = \left(\dfrac{P_4}{P_3}\right)^{(\gamma-1)/\gamma}$ and $T_3 = 850 + 273 = \underline{1\,123 \text{ K}}$

$\therefore\ T_{4'} = T_3\left(\dfrac{P_4}{P_3}\right)^{(\gamma-1)/\gamma} = 1\,123 \times \left(\dfrac{95}{570}\right)^{(1.4-1)/1.4} = \dfrac{1\,123}{1.668} = \underline{673.26 \text{ K}}$

$\therefore\ \eta_t = \dfrac{1\,123 - T_4}{1\,123 - 673.26} = 0.84$

from which,

$T_4 = 1\,123 - 0.84(1\,123 - 673.26)$

$= 1\,123 - (0.84 \times 449.74)$

$= 1\,123 - 377.78$

$= \underline{745.22 \text{ K}}$

$\oint W = \dot{m}c_p\{(T_3 - T_4) - (T_2 - T_1)\}$, (see equation (14) and section 4.13.)

$= 3 \times 1.006 \times \{(1\,123 - 745.22) - (517.17 - 285)\}$

$= 3 \times 1.006 \times (377.78 - 232.17)$

$$= 3 \times 1.006 \times 145.61$$
$$= \underline{439.45 \text{ kJ/s}}$$
$$= \underline{439.45 \text{ kW}}, \quad \text{since } 1 \text{ kJ/s} = 1 \text{ kW}$$

(b) Thermal $\eta = \dfrac{\text{Net cycle work}}{\text{Heat received}}$

$$= \frac{439.45}{\dot{m}c_p(T_3 - T_2)} = \frac{439.45}{3 \times 1.006 \times (1\,123 - 517.17)}$$

$$= \frac{439.45}{3 \times 1.006 \times 605.83}$$

$$= \frac{439.45}{1\,828.39} = \underline{0.24}$$

$$= 0.24 \times 100\% = \underline{24\%}$$

(c) Work ratio $= \dfrac{\text{net cycle work}}{\text{positive cycle work}}$

$$= \frac{439.45}{\dot{m}c_p(T_3 - T_4)} = \frac{439.45}{3 \times 1.006 \times (1\,123 - 745.22)}$$

$$= \frac{439.45}{3 \times 1.006 \times 377.78} = \frac{439.45}{1\,140.14}$$

$$= \underline{0.385}$$

Example 20
A gas turbine operating on the ideal constant pressure cycle has the temperature limits of 700 °C and 20 °C. Compression in the compressor and expansion in the turbine are isentropic. Take $\gamma = 1.4$ and $c_p = 1.005$ kJ/kgK. Determine:
(a) the pressure ratio which will give maximum net work output;
(b) the maximum net specific work output;
(c) the thermal efficiency at maximum work output;
(d) the work ratio at the maximum work output.

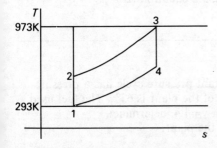

$T_3 = 700 + 273 = \underline{973\text{K}}$
$T_1 = 20 + 273 = \underline{293\text{K}}$

(a) Maximum pressure ratio $= r_{pmax} = \left(\dfrac{T_3}{T_1}\right)^{\gamma/(\gamma-1)}$, see equation (15)

$$= \left(\dfrac{973}{293}\right)^{1.4/(1.4-1)} = 3.321^{1.4/0.4} = 3.321^3$$

$$= \underline{66.75}$$

For maximum network output,

Pressure ratio $r_p = \sqrt{r_{pmax}}$, see equation (22)

$$= \sqrt{66.75}$$

$$= \underline{8.17}$$

(b) $\dfrac{T_2}{T_1} = r_p^{(\gamma-1)/\gamma}$

$\therefore\ T_2 = T_1 r_p^{(\gamma-1)/\gamma} = 293 \times 8.17^{(1.4-1)/1.4} = 293 \times 8.17^{1/3.5} = 293 \times 1.822$

$$= \underline{533.85\ \text{K}}$$

$$= T_4, \quad \text{see equation (23)}$$

Maximum net specific work output

$$= c_p\{(T_3 - T_4) - (T_2 - T_1)\}$$
$$= 1.005\{(973 - 533.85) - (533.85 - 293)\}$$
$$= 1.005\{439.15 - 240.85\}$$
$$= 1.005 \times 198.3$$
$$= \underline{199.3\ \text{kJ}}$$

(c) Thermal $\eta = \dfrac{199.3}{c_p(T_3 - T_2)} = \dfrac{199.3}{1.005(973 - 533.85)}$

$$= \dfrac{199.3}{1.005 \times 439.15} = \dfrac{199.3}{441.35}$$

$$= \underline{0.452} = 0.452 \times 100\% = \underline{45.2\%}$$

(d) Work ratio = Thermal $\eta = \underline{0.452}$, see equation (24)

Example 21
A gas turbine operating on a constant pressure cycle has a pressure ratio of 7:1. The relative efficiency of the plant is 65%. The fuel used has a calorific value of 43 MJ/kg. If $\gamma = 1.4$, determine:
(a) the actual thermal efficiency of the plant;

(b) the specific fuel consumption in kg/kWh.

(a) Ideal thermal $\eta = 1 - \dfrac{1}{r_v^{\gamma-1}}$, where $r_v = \dfrac{V_1}{V_2}$

But the pressure ratio is given $= \dfrac{P_2}{P_1} = 7$

Now,
$$\frac{T_1}{T_2} = \left(\frac{P_1}{P_2}\right)^{(\gamma-1)/\gamma} = \left(\frac{V_2}{V_1}\right)^{\gamma-1} = \frac{1}{r_v^{\gamma-1}}$$

∴ Ideal thermal $\eta = 1 - \left(\dfrac{P_1}{P_2}\right)^{(\gamma-1)/\gamma}$

$$= 1 - \frac{1}{7^{(1.4-1)/1.4}} = 1 - \frac{1}{7^{0.4/1.4}}$$

$$= 1 - \frac{1}{7^{1/3.5}} = 1 - \frac{1}{1.744} = 1 - 0.573$$

$$= \underline{0.427} = 0.427 \times 100\% = \underline{42.7\%}$$

Relative $\eta = \dfrac{\text{actual thermal } \eta}{\text{ideal thermal } \eta}$

∴ Actual thermal $\eta =$ Relative $\times$ ideal thermal η

$= 0.65 \times 0.427$

$= \underline{0.278} = 0.278 \times 100\% = \underline{27.8\%}$

(b) Energy to net work/kg fuel $= 43\,000 \times 0.278$

$= \underline{11\,954 \text{ kJ}}$

Energy equivalent of 1 kWh $= 3\,600$ kJ

∴ Specific fuel consumption $= \dfrac{3\,600}{11\,954} = \underline{0.301 \text{ kg/kWh}}$

5.5 The Diesel cycle

This cycle is so named after Rudolph Diesel (1858–1913) who had German parents but was born in Paris. From about 1890, he pioneered a great deal of work on fuel injection oil engines. He had a considerable interest in the use of coal as a fuel for use in internal combustion engines. As a result of his work, engines using fuel oil are commonly called Diesel engines. Rudolph Diesel mysteriously disappeared while crossing the English Channel in 1913.

The cycle which bears his name is illustrated in Fig. 5.10.

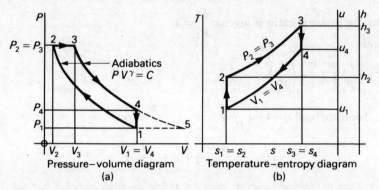

Fig. 5.10

In Fig. 5.10(a) is represented the $P-V$ diagram of the Diesel cycle. In Fig. 5.10(b) is represented the corresponding $T-s$ diagram. The cycle is arranged as follows:

1–2 Adiabatic compression according to the law $PV^\gamma = C$.
   Pressure increases from P_1 to P_2.
   Temperature increases from T_1 to T_2.
   Volume decreases from V_1 to V_2.
   Entropy remains constant at $s_1 = s_2$.

2–3 Constant pressure heat addition.
   Pressure remains constant at $P_2 = P_3$.
   Temperature increases from T_2 to T_3.
   Volume increases from V_2 to V_3.
   Entropy increases from s_2 to s_3.

3–4 Adiabatic expansion according to the law $PV^\gamma = C$.
   Pressure decreases from P_3 to P_4.
   Temperature decreases from T_3 to T_4.
   Volume increases from V_3 to V_4.
   Entropy remains constant at $s_3 = s_4$.

4–1 Constant volume heat rejection.
   Volume remains constant at $V_4 = V_1$.
   Temperature decreases from T_4 to T_1.
   Pressure decreases from P_4 to P_1.
   Entropy decreases from s_4 to s_1.

This process completes the cycle and returns the gas to its orginal state.

This cycle is sometimes referred to as the modified constant pressure cycle. Inspection of Fig. 5.10(a) will show the reason for this. If, instead of cutting off the expansion at 4, the gas was allowed to expand completely to 5, then, in order to return the gas to its original state at 1, constant pressure heat rejection would have to take place from 5 to 1.

This is shown dotted. The diagram 1 2 3 5 is the constant pressure cycle and, by cutting off the part 1 4 5, it is modified into the Diesel cycle.

In practice, by cutting off part 1 4 5 of the cycle a considerable saving in cylinder volume would be obtained. The area 1 4 5 represents a small amount of work which does not really justify the increase of the cylinder volume from V_1 to V_5.

An analysis of the properties at state points 1, 2, 3 and 4 can be made. Again it is here assumed that P_1, V_1 and T_1 are known.

1. P_1, V_1, T_1.
2. Assume that the volume ratio V_1/V_2 is known.

$$T_1/T_2 = (V_2/V_1)^{\gamma-1},$$

$$\therefore \ T_2 = T_1\left(\frac{V_1}{V_2}\right)^{\gamma-1} = T_1 r_v^{\gamma-1}$$

where $r_v = V_1/V_2 =$ adiabatic compression, volume ratio.
Also, $P_1 V_1^\gamma = P_2 V_2^\gamma$,

$$\therefore \ P_2 = P_1\left(\frac{V_1}{V_2}\right)^\gamma = P_1 r_v^\gamma$$

3. $P_3 = P_2$, since the pressure remsins constant.
$V_3/T_3 = V_2/T_2$,

$$\therefore \ T_3 = T_2 \frac{V_3}{V_2} = \frac{V_3}{V_2} T_1 r_v^{\gamma-1} = \beta T_1 r_v^{\gamma-1}$$

where $\beta = V_3/V_2 =$ cut-off ratio.

4. $T_4/T_3 = (V_3/V_4)^{\gamma-1}$,

$$\therefore \ T_4 = T_3\left(\frac{V_3}{V_4}\right)^{\gamma-1}$$

Now $V_3/V_4 = V_3/V_1$, since $V_4 = V_1$, and

$$\frac{V_3}{V_1} = \frac{V_3}{V_2}\frac{V_2}{V_1} = \frac{\beta}{r_v}$$

$$\therefore \ T_4 = T_3\left(\frac{\beta}{r_v}\right)^{\gamma-1} = \beta T_1 r_v^{\gamma-1}\frac{\beta^{\gamma-1}}{r_v^{\gamma-1}}$$

or

$$T_4 = \beta^\gamma T_1$$

Also $P_4 V_4^\gamma = P_3 V_3^\gamma$,

$$\therefore P_4 = P_3 \left(\frac{V_3}{V_4}\right)^\gamma = P_3 \left(\frac{\beta}{r_v}\right)^\gamma$$

The work done during the cycle may be determined as follows:

Processes 2–3 and 3–4 are expansions and hence give positive work done. Process 1–2 is a compression and hence will give negative work done.

The net work done during the cycle will be the sum of the work done during these processes.

Hence,

$$\oint W = \text{Area under 2–3} + \text{area under 3–4} - \text{area under 1–2}$$

$$= P_2(V_3 - V_2) + \frac{(P_3 V_3 - P_4 V_4)}{\gamma - 1} - \frac{(P_2 V_2 - P_1 V_1)}{\gamma - 1}$$

$$= P_2(V_3 - V_2) + \frac{(P_3 V_3 - P_4 V_4) - (P_2 V_2 - P_1 V_1)}{\gamma - 1} \tag{1}$$

$$= mR(T_3 - T_2) + \frac{mR}{\gamma - 1}\{(T_3 - T_4) - (T_2 - T_1)\}$$

$$= mR\left\{(T_3 - T_2) + \frac{(T_3 - T_4) - (T_2 - T_1)}{\gamma - 1}\right\} \tag{2}$$

Alternatively, the work done may be obtained from the equation,

Word done = Heat received − Heat rejected.

In this cycle, heat is received during constant pressure process 2–3 and rejected during constant volume process 4–1.

No heat is received or rejected during the adiabatic processes.

Hence,

$$\oint W = mc_p(T_3 - T_2) - mc_v(T_4 - T_1) \tag{3}$$

Alternatively,

$$\oint W = \text{Heat received} \times \text{Thermal } \eta$$

or

$$\oint W = mc_p(T_3 - T_2) \times \text{Thermal } \eta \tag{4}$$

The thermal efficiency may be determined as follows:

Thermal $\eta = 1 - \dfrac{\text{Heat rejected}}{\text{Heat received}}$

$$= 1 - \frac{mc_v(T_4 - T_1)}{mc_p(T_3 - T_2)}$$

$$= 1 - \frac{1}{\gamma}\frac{(T_4 - T_1)}{(T_3 - T_2)} \tag{5}$$

This gives the thermal efficiency in terms of temperatures.

Also, substituting temperatures in terms of T_1 into equation (5), then,

Thermal $\eta = 1 - \dfrac{1}{\gamma}\dfrac{(\beta^\gamma T_1 - T_1)}{(\beta T_1 r_v^{\gamma-1} - T_1 r_v^{\gamma-1})}$

$$= 1 - \frac{1}{\gamma}\frac{T_1(\beta^\gamma - 1)}{T_1 r_v^{\gamma-1}(\beta - 1)}$$

$$= 1 - \frac{1}{r_v^{\gamma-1}}\frac{1}{\gamma}\frac{(\beta^\gamma - 1)}{(\beta - 1)} \tag{6}$$

Now, for a constant pressure process,

Heat received or rejected = Change of enthalpy

For a constant volume process,

Heat received or rejected = Change of internal energy.

∴ Using the T–s chart,

Heat received = $h_3 - h_2$

Heat rejected = $u_4 - u_1$

Thermal $\eta = 1 - \dfrac{\text{Heat rejected}}{\text{Heat received}}$

$$= 1 - \frac{(u_4 - u_1)}{(h_3 - h_2)} \tag{7}$$

Also,

$\oint W =$ Heat received − Heat rejected

$$= (h_3 - h_2) - (u_4 - u_1) \tag{8}$$

Once again this gives the work done/unit mass/cycle, since the chart will be made out for unit mass of gas.

Example 22

At the beginning of the adiabatic compression of an ideal Diesel cycle, using air, the pressure, volume and temperature are 95 kN/m², 0.01 m³ and 30 °C, respectively. The overall volume ratio of the cycle is 12:1. The maximum temperature of the cycle is 1 350 °C.
 Take, $\gamma = 1.4$ and $c_v = 0.717$ kJ/kgK.
 Determine:
(a) the pressure, volume and temperature at the cycle change points;
(b) the net work done/cycle;
(c) the ideal thermal efficiency of the cycle;
(d) the work ratio of the cycle;
(e) the mean effective pressure of the cycle.

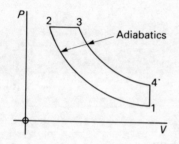

(a) $P_1 = 95$ kN/m², $V_1 = 0.01$ m³, $t_1 = 30$ °C, given.

$$\frac{T_2}{T_1} = \left(\frac{V_1}{V_2}\right)^{\gamma-1} \quad \text{and} \quad T_1 = 30 + 273 = \underline{303 \text{ K}}$$

$$\therefore T_2 = T_1\left(\frac{V_1}{V_2}\right)^{\gamma-1} = 303 \times 12^{1.4-1} = 303 \times 12^{0.4} = 303 \times 2.7 = \underline{818.1 \text{ K}}$$

$$t_2 = 818.1 - 273 = \underline{545.1 \text{ °C}}$$

$$V_2 = \frac{V_1}{12} = \frac{0.01}{12} = \underline{0.000\,83 \text{ m}^3}$$

$$P_1 V_1^\gamma = P_2 V_2^\gamma$$

$$\therefore P_2 = P_1\left(\frac{V_1}{V_2}\right)^\gamma = 95 \times 12^{1.4} = 95 \times 32.42$$

$$= 3\,080 \text{ kN/m}^2 = \underline{3.08 \text{ MN/m}^2}$$

$$P_3 = P_2 = \underline{3.08 \text{ MN/m}^2}$$

$$\frac{V_2}{T_2} = \frac{V_3}{T_3} \quad \text{and} \quad T_3 = 1\,350 + 273 = 1\,623 \text{ K, given.}$$

$$\therefore V_3 = V_2 \frac{T_3}{T_2} = 0.00083 \times \frac{1623}{818.1}$$

$$= \underline{0.00165 \text{ m}^3}$$

$$V_4 = V_1 = \underline{0.01 \text{ m}^3}$$

$$P_3 V_3^\gamma = P_4 V_4^\gamma$$

$$\therefore P_4 = P_3 \left(\frac{V_3}{V_4}\right)^\gamma = 3080 \times \left(\frac{0.00165}{0.01}\right)^{1.4} = \frac{3080}{6.06^{1.4}} = \frac{3080}{12.46}$$

$$= \underline{247.2 \text{ kN/m}^2}$$

$$\frac{P_4}{T_4} = \frac{P_1}{T_1}$$

$$\therefore T_4 = T_1 \frac{P_4}{P_1} = 303 \times \frac{247.2}{95}$$

$$= \underline{788.4 \text{ K}}$$

$$t_4 = 788.4 - 273$$

$$= \underline{515.4 \text{ °C}}$$

$$V_4 = V_1 = \underline{0.01 \text{ m}^3}$$

Hence,

$P_1 = 95$ kN/m²	$V_1 = 0.01$ m³	$t_1 = 30$ °C
$P_2 = 3.08$ MN/m²	$V_2 = 0.00083$ m³	$t_2 = 545.1$ °C
$P_3 = 3.08$ MN/m²	$V_3 = 0.00165$ m³	$t_3 = 1350$ °C
$P_4 = 247.2$ kN/m²	$V_4 = 0.01$ m³	$t_4 = 515.4$ °C

(b) $\oint W = m c_p (T_3 - T_2) - m c_v (T_4 - T_1)$, see equation (3)

Now $\dfrac{c_p}{c_v} = \gamma \quad \therefore c_p = \gamma c_v = 1.4 \times 0.717 = \underline{1.0038 \text{ kJ/kgK}}$

$R = c_p - c_v = 1.0038 - 0.717 = \underline{0.2868 \text{ kJ/kgK}}$

$P_1 V_1 = m R T_1$

$\therefore m = \dfrac{P_1 V_1}{R T_1} = \dfrac{95 \times 0.01}{0.2868 \times 303}$

$= \underline{0.011 \text{ kg}}$

Hence,

$$\oint W = 0.011 \times 1.0038 (1623 - 818.1) - 0.011 \times 0.717 (788.4 - 303)$$

$$= (0.011 \times 1.0038 \times 804.9) - (0.011 \times 0.717 \times 485.4)$$

$$= 8.89 - 3.83$$
$$= \underline{5.06 \text{ kJ}}$$

(c) Thermal $\eta = \dfrac{\text{net work done}}{\text{heat received}}$

$$= \dfrac{5.06}{mc_p(T_3 - T_2)} = \dfrac{5.06}{0.011 \times 1.003\,8 \times (1\,623 - 818.1)}$$

$$= \dfrac{5.06}{0.011 \times 1.003\,8 \times 804.9}$$

$$= \dfrac{5.06}{0.89} = \underline{0.57}$$

$$= 0.57 \times 100\% = \underline{57\%}$$

(d) Work ratio $= \dfrac{5.06}{P_2(V_3 - V_2) + \dfrac{(P_3 V_3 - P_4 V_4)}{\gamma - 1}}$

$$= \dfrac{5.06}{3\,080(0.001\,65 - 0.000\,83) + \dfrac{(3\,080 \times 0.001\,65 - 247.2 \times 0.01)}{1.4 - 1}}$$

$$= \dfrac{5.06}{(3080 \times 0.000\,82) + \dfrac{(5.082 - 2.472)}{0.4}}$$

$$= \dfrac{5.06}{2.526 + \dfrac{2.61}{0.4}} = \dfrac{5.06}{2.526 + 6.525}$$

$$= \dfrac{5.06}{9.051}$$

$$= \underline{0.56}$$

(e) Mean effective pressure $= \dfrac{5.06}{V_1 - V_2} = \dfrac{5.06}{0.01 - 0.000\,83}$

$$= \dfrac{5.06}{0.009\,17}$$

$$= \underline{551.8 \text{ kN/m}^2}$$

Example 23

An oil engine works on the ideal Diesel cycle. The overall volume ratio of the cycle is 13:1 and constant pressure heat addition ceases at 0.15

of the stroke. The pressure at the commencement of compression is $100\,\text{kN/m}^2$. The engine uses air at the rate of $0.1\,\text{m}^3/\text{s}$ at the initial intake conditions. If $\gamma = 1.395$, determine:
(a) the thermal efficiency of the cycle;
(b) the net power output of the engine.

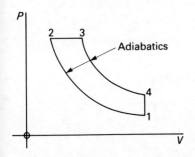

(a) Stroke volume $= 13 - 1 = \underline{12 \text{ volumes}}$

0.15 of stroke volume $= 0.15 \times 12 = \underline{1.8 \text{ volumes}}$

∴ constant pressure heat addition ceases at $1 + 1.8 = \underline{2.8 \text{ volumes}}$

Thermal $\eta = 1 - \dfrac{1}{r_v^{\gamma-1}} \dfrac{1}{\gamma} \dfrac{(\beta^\gamma - 1)}{(\beta - 1)}$, see equation (6)

and $r_v = 13$, $\beta = \dfrac{V_3}{V_2} = \dfrac{2.8}{1} = 2.8$

∴ Thermal $\eta = 1 - \dfrac{1}{13^{0.395}} \times \dfrac{1}{1.395} \times \dfrac{(2.8^{1.395} - 1)}{(2.8 - 1)}$

$= 1 - \dfrac{1}{2.754} \times \dfrac{1}{1.395} \times \left(\dfrac{4.205 - 1}{1.8} \right)$

$= 1 - \dfrac{1}{2.754 \times 1.395} \times \dfrac{3.205}{1.8}$

$= 1 - 0.463$

$= \underline{0.537}$

$= 0.537 \times 100\% = \underline{53.7\%}$

(b) Let $V_1 - V_2 = 0.1\,\text{m}^3$

Also, $\dfrac{V_1}{V_2} = \dfrac{13}{1}$

hence $\dfrac{V_1}{V_2} - 1 = \dfrac{13}{1} - 1$

or, $\dfrac{V_1 - V_2}{V_2} = 12$

from which, $\dfrac{0.1}{V_2} = 12$

hence, $V_2 = \dfrac{0.1}{12}$

$\qquad = \underline{0.008\ 3\ \text{m}^3}$

$\therefore\ V_1 = 0.1 + 0.008\ 3 = \underline{0.108\ 3\ \text{m}^3}$

$V_3 = 0.008\ 3 + (0.1 \times 0.15) = 0.008\ 3 + 0.015$

$\qquad = \underline{0.023\ 3\ \text{m}^3}$

$V_4 = V_1 = \underline{0.108\ 3\ \text{m}^3}$

Now,

$P_1 V_1^\gamma = P_2 V_2^\gamma$

$\therefore\ P_2 = P_1 \left(\dfrac{V_1}{V_2}\right)^\gamma = 100 \times 13^{1.395} = 100 \times 35.8$

$\qquad = 3\ 580\ \text{kN/m}^2 = \underline{3.58\ \text{MN/m}^2}$

$P_3 = P_2 = \underline{3.58\ \text{MN/m}^2}$

$P_4 V_4^\gamma = P_3 V_3^\gamma$

$\therefore\ P_4 = P_3 \left(\dfrac{V_3}{V_4}\right)^\gamma = 3\ 580 \times \left(\dfrac{0.023\ 3}{0.108\ 3}\right)^{1.395}$

$\qquad = \dfrac{3\ 580}{4.65^{1.395}} = \dfrac{3\ 580}{8.53}$

$\qquad = \underline{419.7\ \text{kN/m}^2}$

$\oint W/s = P_2(V_3 - V_2) + \dfrac{\{(P_3 V_3 - P_4 V_4) - (P_2 V_2 - P_1 V_1)\}}{\gamma - 1}$

$= 3\ 580(0.023\ 3 - 0.008\ 3)$

$+ \dfrac{\{(3\ 580 \times 0.023\ 3 - 419.7 \times 0.108\ 3) - (3\ 580 \times 0.008\ 3 - 100 \times 0.108\ 3)\}}{1.395 - 1}$

$= (3\ 580 \times 0.015) + \dfrac{\{(83.4 - 45.45) - (29.7 - 10.83)\}}{0.395}$

$= 53.7 + \dfrac{(37.95 - 18.87)}{0.395} = 53.7 + \dfrac{17.08}{0.395}$

$= 53.7 + 43.2$

$= \underline{96.9\ kJ/s}$

$= \underline{96.9\ kW}$

5.6 The dual combustion cycle

This cycle is sometimes referred to as the *composite* cycle.

In this cycle heat is received partly at constant volume and partly at constant pressure, hence the name, dual combustion cycle. Heat is rejected at constant volume.

This cycle fits the actual cycle of an oil engine rather more closely than the Diesel cycle.

This is because in an actual oil engine part of the fuel burns very nearly at constant volume followed by the remainder of the fuel burning very nearly at constant pressure.

The cycle is sometimes also referred to as the high-speed Diesel cycle.

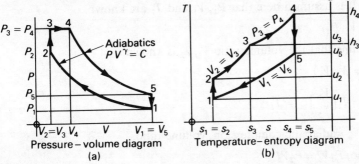

Fig. 5.11

In Fig. 5.11(a) is represented a P–V diagram of the dual combustion cycle. In Fig. 5.11(b) is represented the corresponding T–s diagram. The cycle is arranged as follows:

1–2 Adiabatic compression according to the law $PV^\gamma = C$.
　　Pressure increases from P_1 to P_2.
　　Temperature increases from T_1 to T_2.
　　Volume decreases from V_1 to V_2.
　　Entropy remains constant at $s_1 = s_2$.

2–3 Constant volume heat addition.
　　Volume remains constant at $V_2 = V_3$.
　　Pressure increases from P_2 to P_3.
　　Temperature increases from T_2 to T_3.
　　Entropy increases from s_2 to s_3.

3–4 Constant pressure heat addition.
Pressure remains constant at $P_3 = P_4$.
Volume increases from V_3 to V_4.
Temperature increases from T_3 to T_4.
Entropy increases from s_3 to s_4.

4–5 Adiabatic expansion according to the law $PV^\gamma = C$.
Pressure decreases from P_4 to P_5.
Temperature decreases from T_4 to T_5.
Volume increases from V_4 to V_5.
Entropy remains constant at $s_4 = s_5$.

5–1 Constant volume heat rejection.
Volume remains constant at $V_5 = V_1$.
Pressure decreases from P_5 to P_1.
Temperature decreases from T_5 to T_1.
Entropy decreases from s_5 to s_1.

This process completes the cycle and returns the gas to its original state.

An analysis of the properties at state points 1, 2, 3, 4 and 5 can be made. It is assumed here that P_1, V_1 and T_1 are known.

1. P_1, V_1, T_1.

2. Assume that the volume ratio V_1/V_2 is known.

$T_1/T_2 = (V_2/V_1)^{\gamma-1}$,

$$\therefore\ T_2 = T_1\left(\frac{V_1}{V_2}\right)^{\gamma-1} = T_1 r_v^{\gamma-1}$$

where r_v = adiabatic compression, volume ratio.

Also, $P_1 V_1^\gamma = P_2 V_2^\gamma$,

$$\therefore\ P_2 = P_1\left(\frac{V_1}{V_2}\right)^\gamma = P_1 r_v^\gamma$$

3. $P_3/T_3 = P_2/T_2$, since the volume remains constant at $V_2 = V_3$,

$$\therefore\ T_3 = \frac{P_3}{P_2} T_2 = \alpha T_1 r_v^{\gamma-1}$$

where $\alpha = P_3/P_2$ = constant volume heat addition pressure ratio.

4. $V_3/T_3 = V_4/T_4$, since the pressure remains constant at $P_3 = P_4$,

$$\therefore\ T_4 = \frac{V_4}{V_3} T_3 = \alpha\beta T_1 r_v^{\gamma-1}$$

where $\beta = V_4/V_3$ = constant pressure heat addition cut-off ratio.

5. $T_5/T_4 = (V_4/V_5)^{\gamma-1}$,

$$\therefore\ T_5 = T_4\left(\frac{V_4}{V_5}\right)^{\gamma-1}$$

Now $V_5 = V_1$,

$$\therefore \frac{V_4}{V_5} = \frac{V_4}{V_1} = \frac{V_4}{V_3}\frac{V_3}{V_1}$$

Also $V_3 = V_2$,

$$\therefore \frac{V_4}{V_5} = \frac{V_4}{V_3}\frac{V_2}{V_1} = \frac{\beta}{r_v}$$

since $V_4/V_3 = \beta$ and $V_1/V_2 = r_v$

∴ from above,

$$T_3 = \alpha\beta T_1 r_v^{\gamma-1}\frac{\beta^{\gamma-1}}{r_v^{\gamma-1}} = \alpha\beta^\gamma T_1$$

Note also that for the constant volume process 5–1, $P_5/T_5 = P_1/T_1$,

$$\therefore T_5 = \frac{P_5}{P_1} T_1$$

It appears from this then that,

$$\frac{P_5}{P_1} = \alpha\beta^\gamma$$

Now this can be shown as follows:

$$\alpha\beta^\gamma = \frac{P_3}{P_2}\left(\frac{V_4}{V_3}\right)^\gamma$$

$P_3 V_4^\gamma = P_4 V_4^\gamma$ since $P_3 = P_4$

and

$P_4 V_4^\gamma = P_5 V_5^\gamma$ since for the process 4–5, $PV^\gamma = C$

Also,

$P_2 V_3^\gamma = P_2 V_2^\gamma$ since $V_2 = V_3$

and $P_2 V_2^\gamma = P_1 V_1^\gamma$ since for the process 1–2, $PV^\gamma = C$

$$\therefore \alpha\beta^\gamma = \frac{P_5 V_5^\gamma}{P_1 V_1^\gamma} = \frac{P_5}{P_1} \quad \text{since} \quad V_1 = V_5$$

Note also that, $P_5 V_5^\gamma = P_4 V_4^\gamma$,

$$\therefore P_5 = P_4\left(\frac{V_4}{V_5}\right)^\gamma = P_4\left(\frac{\beta}{r_v}\right)^\gamma$$

The work done during the cycle may be obtained as follows:

Processes 3–4 and 4–5 are expansions and hence give positive work done.

Process 1–2 is a compression and hence gives negative work done.

The net work done during the cycle will be the sum of the work done during these processes.

Hence,

$\oint W =$ Area under 3–4 + area under 4–5 – area under 1–2

$$= P_3(V_4 - V_3) + \frac{(P_4 V_4 - P_5 V_5)}{\gamma - 1} - \frac{(P_2 V_2 - P_1 V_1)}{\gamma - 1}$$

$$= P_3(V_4 - V_3) + \frac{(P_4 V_4 - P_5 V_5) - (P_2 V_2 - P_1 V_1)}{\gamma - 1} \tag{1}$$

$$= mR(T_4 - T_3) + \frac{mR}{\gamma - 1}\{(T_4 - T_5) - (T_2 - T_1)\}$$

$$= mR\left\{(T_4 - T_3) + \frac{(T_4 - T_5) - (T_2 - T_1)}{\gamma - 1}\right\} \tag{2}$$

Alternatively, the work done may be obtained from the equation.

Work done = Heat received – Heat rejected

In this cycle heat is received during constant volume process 2–3 and constant pressure process 3–4 and is rejected during constant volume process 5–1.

No heat is received or rejected during the adiabatic processes.

$$\therefore \oint W = mc_v(T_3 - T_2) + mc_p(T_4 - T_3) - mc_v(T_5 - T_1) \tag{3}$$

Alternatively,

$\oint W =$ Heat received $\times$ Thermal η

$$= \{mc_v(T_3 - T_2) + mc_p(T_4 - T_3)\} \times \text{Thermal } \eta \tag{4}$$

The thermal efficiency may be determined as follows:

Thermal $\eta = 1 - \dfrac{\text{Heat rejected}}{\text{Heat received}}$

$$= 1 - \frac{mc_v(T_5 - T_1)}{mc_v(T_3 - T_2) + mc_p(T_4 - T_3)}$$

$$= 1 - \frac{(T_5 - T_1)}{(T_3 - T_2) + \gamma(T_4 - T_3)} \tag{5}$$

This gives the thermal efficiency in terms of temperatures.

Also, substituting temperatures in terms of T_1 into equation (5), then,

Thermal $\eta = 1 - \dfrac{(\alpha\beta^\gamma T_1 - T_1)}{(\alpha T_1 r_v^{\gamma-1} - T_1 r_v^{\gamma-1}) + \gamma(\alpha\beta T_1 r_v^{\gamma-1} - \alpha T_1 r_v^{\gamma-1})}$

$= 1 - \dfrac{T_1(\alpha\beta^\gamma - 1)}{T_1 r_v^{\gamma-1}\{(\alpha-1) + \alpha\gamma(\beta-1)\}}$

$= 1 - \dfrac{1}{r_v^{\gamma-1}} \left\{ \dfrac{(\alpha\beta^\gamma - 1)}{(\alpha-1) + \alpha\gamma(\beta-1)} \right\}$ (6)

Note that this expression contains the expressions for thermal efficiency of both the constant volume and the Diesel cycles.

If $\beta = 1$, then there is no constant pressure heat addition since $V_4 = V_3$ and substituting this into equation (6)

Thermal $\eta = 1 - \dfrac{1}{r_v^{\gamma-1}}$

i.e. the constant volume cycle.

If $\alpha = 1$, then there is no constant volume heat addition, since $P_3 = P_2$ and substituting this into equation (6),

Thermal $\eta = 1 - \dfrac{1}{r_v^{\gamma-1}} \dfrac{1}{\gamma} \dfrac{(\beta^\gamma - 1)}{(\beta - 1)}$

i.e. the Diesel cycle.

For the dual combustion cycle, using the T–s chart,

Heat received at constant volume $= u_3 - u_2$

Heat received at constant pressure $= h_4 - h_3$

Heat rejected at constant volume $= u_5 - u_1$

Thermal $\eta = 1 - \dfrac{\text{Heat rejected}}{\text{Heat received}}$

$= 1 - \dfrac{(u_5 - u_1)}{(u_3 - u_2) + (h_4 - h_3)}$ (7)

$\oint W = (u_3 - u_2) + (h_4 - h_3) - (u_5 - u_1)$ (8)

Again, this gives the work done/unit mass/cycle, since the chart will be made out for unit mass of gas.

Example 24
An ideal dual combustion cycle has an overall volume ratio of 12:1. The pressure, volume and temperature at the commencement of the adiabatic compression are 120 kN/m^2, 0.12 m^3 and $35\,°\text{C}$, respectively. The maximum pressure of the cycle is 5 MN/m^2 and the maximum temperature of the cycle is $1\,250\,°\text{C}$.

Take, $c_p = 1.006$ kJ/kgK, $c_v = 0.717$ kJ/kgK. Determine:

(a) the pressure, volume and temperature at the cycle change points;
(b) the net work done/cycle;
(c) the thermal efficiency of the cycle;
(d) the work ratio of the cycle;
(e) the mean effective pressure of the cycle.

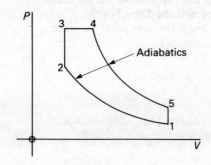

(a) $P_1 = 120$ kN/m², $V_1 = 0.12$ m³, $t_1 = 35$ °C, given.

$$V_2 = \frac{V_1}{12} = \frac{0.12}{12} = \underline{0.01 \text{ m}^3}$$

$$\gamma = \frac{c_p}{c_v} = \frac{1.006}{0.717} = \underline{1.403}$$

$$P_1 V_1^\gamma = P_2 V_2^\gamma$$

$$\therefore P_2 = P_1 \left(\frac{V_1}{V_2}\right)^\gamma = 120 \times 12^{1.403} = 120 \times 32.7 = \underline{3\,924 \text{ kN/m}^2}$$

$$= \underline{3.924 \text{ MN/m}^2}$$

$$\frac{T_2}{T_1} = \left(\frac{V_1}{V_2}\right)^{\gamma-1}$$

$$\therefore T_2 = T_1 \left(\frac{V_1}{V_2}\right)^{\gamma-1} \quad \text{and} \quad T_1 = 35 + 273 = \underline{308 \text{ K}}$$

$$= 308 \times 12^{1.403-1} = 308 \times 12^{0.403} = 308 \times 2.72 = \underline{837.8 \text{ K}}$$

$$t_2 = 837.8 - 273 = \underline{564.8 \text{ °C}}$$

$$V_3 = V_2 = \underline{0.01 \text{ m}^3}$$

$$\frac{P_3}{T_3} = \frac{P_2}{T_2}$$

$\therefore \ T_3 = T_2 \dfrac{P_3}{P_2}$ and $P_3 = \underline{5 \text{ MN/m}^2}$, given.

$$= 837.8 \times \frac{5}{3.924} = \underline{1\,067.5 \text{ K}}$$

$t_3 = 1\,067.5 - 273 = \underline{794.5 \text{ °C}}$

$$\frac{V_4}{T_4} = \frac{V_3}{T_3}$$

$\therefore \ V_4 = V_3 \dfrac{T_4}{T_3}$ and $t_4 = \underline{1\,250 \text{ °C}}$, given

$$T_4 = 1\,250 + 273 = \underline{1\,523 \text{ K}}$$

$$= 0.01 \times \frac{1\,523}{1\,067.5} = \underline{0.014\,3 \text{ m}^3}$$

$P_4 = P_3 = \underline{5 \text{ MN/m}^2}$

$V_5 = V_1 = \underline{0.12 \text{ m}^3}$

$P_4 V_4^\gamma = P_5 V_5^\gamma$

$$\therefore \ P_5 = P_4 \left(\frac{V_4}{V_5}\right)^\gamma = 5\,000 \times \left(\frac{0.014\,3}{0.12}\right)^{1.403} = \frac{5\,000}{8.39^{1.403}}$$

$$= \frac{5\,000}{19.77} = \underline{252.9 \text{ kN/m}^2}$$

$$\frac{T_5}{T_4} = \left(\frac{V_4}{V_5}\right)^{\gamma-1}$$

$$\therefore \ T_5 = T_4 \left(\frac{V_4}{V_5}\right)^{\gamma-1} = 1\,523 \times \left(\frac{0.014\,3}{0.12}\right)^{1.403-1} = \frac{1\,523}{8.39^{0.403}}$$

$$= \frac{1\,523}{2.36} = \underline{645.3 \text{ K}}$$

$t_5 = 645.3 - 273 = \underline{372.3 \text{ °C}}$

Hence,

$P_1 = 120 \text{ kN/m}^2 \quad V_1 = 0.12 \text{ m}^3 \quad t_1 = 35 \text{ °C}$
$P_2 = 3.924 \text{ MN/m}^2 \quad V_2 = 0.01 \text{ m}^3 \quad t_2 = 564.8 \text{ °C}$
$P_3 = 5 \text{ MN/m}^2 \quad V_3 = 0.01 \text{ m}^3 \quad t_3 = 794.5 \text{ °C}$
$P_4 = 5 \text{ MN/m}^2 \quad V_4 = 0.014\,3 \text{ m}^3 \quad t_4 = 1\,250 \text{ °C}$
$P_5 = 252.9 \text{ kN/m}^2 \quad V_5 = 0.12 \text{ m}^2 \quad t_5 = 372.3 \text{ °C}$

(b) $R = c_p - c_v = 1.006 - 0.717 = \underline{0.289 \text{ kJ/kgK}}$

$P_1 V_1 = m R T_1$

$\therefore m = \dfrac{P_1 V_1}{R T_1} = \dfrac{120 \times 0.12}{0.289 \times 308} = \underline{0.162 \text{ kg}}$

$\oint W = m\{c_p(T_4 - T_3) + c_v(T_3 - T_2) - c_v(T_5 - T_1)\}$

$= 0.162\{1.006(1\,523 - 1\,067.5) + 0.717(1\,067.5 - 837.8) - 0.717(645.3 - 308)\}$

$= 0.162\{(1.006 \times 455.5) + (0.717 \times 229.7) - (0.717 \times 337.3)\}$

$= 0.162\{458.2 + 164.7 - 241.8\} = 0.162 \times 381.1$

$= \underline{61.7 \text{ kJ}}$

(c) Thermal $\eta = \dfrac{\text{net work done}}{\text{heat received}}$

$= \dfrac{61.7}{m\{c_p(T_4 - T_3) + c_v(T_3 - T_2)\}}$

$= \dfrac{61.7}{0.162\{1.006(1\,523 - 1\,067.5) + 0.717(1\,067.5 - 837.8)\}}$

$= \dfrac{61.7}{0.162\{458.2 + 164.7\}} = \dfrac{61.7}{0.162 \times 622.9} = \dfrac{61.7}{100.9}$

$= \underline{0.611}$

$= 0.611 \times 100\% = \underline{61.1\%}$

(d) Work ratio $= \dfrac{61.7}{P_3(V_4 - V_3) + \dfrac{(P_4 V_4 - P_5 V_5)}{\gamma - 1}}$

$= \dfrac{61.7}{5\,000(0.014\,3 - 0.01) + \dfrac{(5\,000 \times 0.014\,3 - 252.9 \times 0.12)}{1.403 - 1}}$

$= \dfrac{61.7}{(5\,000 \times 0.004\,3) + \dfrac{(71.5 - 30.3)}{0.403}}$

$= \dfrac{61.7}{21.5 + \dfrac{41.2}{0.403}}$

$$= \frac{61.7}{21.5 + 102.2} = \frac{61.7}{123.7}$$
$$= \underline{0.499}$$

(e) Mean effective pressure $= \dfrac{61.7}{V_1 - V_2} = \dfrac{61.7}{0.12 - 0.01} = \dfrac{61.7}{0.11}$

$$= \underline{560.9 \text{ kN/m}^2}$$

5.7 The Stirling cycle

The Stirling cycle is so named after Dr Robert Stirling, a church minister, and his brother James Stirling, an engineer, who, as early as 1816 obtained a British patent on an air engine. In 1845, one such engine was used in a Dundee foundry. The air was used in a closed system and heat was supplied by a furnace through a heating surface. In the system used, the heat transfer into the air was slow and eventually the heating surface burned out causing the abandonment of the engine.

In Figs 5.12(a) and (b) are shown the P–V and T–s diagrams of the Stirling cycle and it will be seen that it is composed of two isothermal processes and two constant volume processes.

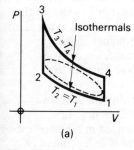
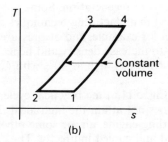

(a) (b)

Fig. 5.12

The importance of this cycle, here, is in the process of regeneration which the Stirling brothers used on their engines. It will be noticed that the two constant volume processes, 2–3 and 4–1, are bounded by the isothermals and therefore have the same temperature limits. Thus, neglecting change in specific heat, the heat transfer required for constant volume process 2–3 is equal to the heat rejection for constant volume process 4–1, or,

$$mc_v(T_3 - T_2) = mc_v(T_4 - T_1)$$

Now, during the process 4–1, this amount of energy is rejected while during process 2–3, the same amount of energy is required. Thus, if a device could be developed within the engine, to store the energy during the rejection of process 4–1 and then subsequently give it up again during process 2–3, when once the necessary first warm-up of the engine had taken place and temperature limits had been established, these two processes would be self perpetuating and no subsequent energy transfer would be required from an external source through the system boundary. The device used to carry out this process was called a *regenerator* and the process called *regeneration*. In the case of the Stirling engine, the regenerator consisted of a matrix of sheet iron plates maintained at the high temperature at one end by the furnace and the low temperature at the other end by a water cooler. Thus the necessary temperature gradient was maintained through the matrix, which was of such a bulk that the necessary energy transfer during the successive processes did not substantially modify the temperatures. Since, with the regenerator installed, external heat transfer is not required in order to carry out the constant volume processes then the thermal efficiency relies only on the two isothermal processes, which is the same as for the Carnot cycle. Hence, with the process of regeneration included, the theoretical Stirling cycle thermal efficiency has the highest thermal efficiency possible $=(T_3 - T_1)/T_3$.

Note that by including the regeneration process, a cycle which would have been otherwise irreversible has been made reversible with its now consequent highest ideal thermal efficiency.

From time to time there is a revival of interest in the Stirling cycle with its process of regeneration. Some modern research has shown that with improved manufacturing techniques and by using gases other than air, hydrogen for example, and at pressures much higher than atmospheric to increase density and hence heat transfer properties, thermal efficiencies as high as 40% with good power outputs are possible.

Note in Fig. 5.11(a) the practical cycle, shown dotted, is an oval shape and does not attain the high ideal thermal efficiency. In a practical Stirling engine, air, or some other gas, is used in a closed cycle and is heated and cooled indirectly. Thus the cycle is independent of the type of fuel as energy source used. Thus nuclear fuel and solar radiation have been considered as possible energy sources as well as the more conventional fuels such as coal, oil, wood, etc.

A further point of interest is that the reversed Stirling cycle has been successfully applied to the refrigeration process required for the liquefaction of air.

For the ideal Stirling cycle,

$$P_3 V_3 = P_4 V_4 \tag{1}$$

$$T_3 = T_4 \tag{2}$$

$$P_2 V_2 = P_1 V_1 \tag{3}$$

$$\oint W = P_3 V_3 \ln \frac{V_4}{V_3} - P_1 V_1 \ln \frac{V_1}{V_2} \qquad (4)$$

and,

$$T_2 = T_1 \qquad (5)$$

$$\frac{V_4}{V_3} = \frac{V_1}{V_2} \qquad (6)$$

$$\therefore \oint W = \ln \frac{V_1}{V_2} (P_3 V_3 - P_1 V_1) \qquad (7)$$

$$= mR \ln \frac{V_1}{V_2} (T_3 - T_1) \quad \text{since } PV = mRT \qquad (8)$$

Ideal thermal $\eta = \dfrac{T_3 - T_1}{T_3} = 1 - \dfrac{T_1}{T_3}$ \qquad (9)

(with regenerator)

$\qquad = $ Carnot η.

Example 25

An ideal Stirling cycle, including regeneration, uses hydrogen. The initial pressure, volume and temperature are 5 MN/m², 0.15 m³ and 35 °C, respectively. The minimum volume of the cycle is 0.06 m³ and the maximum temperature of the cycle is 600 °C. Take $R = 4.12$ kJ/kgK. Determine:

(a) the net work done/cycle;
(b) the ideal thermal efficiency of the cycle.

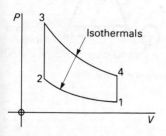

(a) $\oint W = mR \ln \dfrac{V_1}{V_2} (T_3 - T_1)$ and $T_3 = 600 + 273 = \underline{873 \text{ K}}$

$\qquad\qquad\qquad\qquad\qquad\qquad T_1 = 35 + 273 = \underline{308 \text{ K}}$

and $m = \dfrac{P_1 V_1}{R T_1} = \dfrac{5000 \times 0.15}{4.12 \times 308} = \underline{0.59 \text{ kg}}$

$$\therefore \oint W = 0.59 \times 4.12 \times \ln \frac{0.15}{0.06}(873 - 308)$$

$$= 0.59 \times 4.12 \times \ln 2.5 \times 565$$

$$= 0.59 \times 4.12 \times 0.916 \times 565$$

$$= \underline{1\,258 \text{ kJ}}$$

$$= \underline{1.258 \text{ MJ}}$$

(b) Ideal thermal $\eta = 1 - \dfrac{T_1}{T_3}$

$$= 1 - \frac{308}{873} = 1 - 0.353$$

$$= \underline{0.647}$$

$$= 0.647 \times 100\% = \underline{64.7\%}$$

5.8 Steam plant cycles

Before discussing the ideal steam plant cycles it is necessary to explain the general arrangement of the basic elements of a steam plant. These are illustrated diagrammatically in Fig. 5.13.

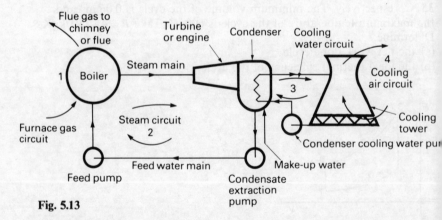

Fig. 5.13

Steam is generated in a *boiler* from which it passes into the steam main. The steam main feeds the steam into a turbine or engine or it may pass into some other plant such as heaters or process machinery. After expending through the turbine or engine or passing through some other plant, if the plant is working on a 'dead-loss' system, then the exhaust steam passes away to atmosphere. Such is the case with the steam locomotive. This system is very inefficient and is rarely adopted in modern plant. It is used in the steam locomotive since, in this case, the plant is mobile and there is not sufficient room for the complex steam recovery equipment which can be installed in a power station or factory. If steam recovery plant is installed then the exhaust steam passes into a *condenser* where it is condensed to water, called *condensate*.

The condensate is extracted from the condenser by the *condensate extraction pump* from which it passes as feed water into the feed water main and back to the boiler. Since the boiler is operating at a high pressure the water pressure must be increased in order to get it into the boiler. This is dealt with by means of a pump called the *feed pump*. Thus the water returns to the boiler and it will be noted that, neglecting losses in the system, with a steam recovery plant, it is the same water being circulated all the time. Actually, there are losses and this is made up in the condenser by means of a make-up water supply. The advantages of steam recovery plant are primarily as follows.

Firstly, the pressure in the condenser can be operated well below atmospheric pressure. This means that a greater expansion of the steam can be obtained which results in more work. Secondly, the water in the circuit can be treated to reduce scale formation in the boiler. The formation of scale in the boiler impedes the transfer of heat from the furnace to the water and hence results in a reduction of boiler efficiency. It may further result in local overheating with resultant damage and it may even cause a burst in the vicinity if overheating is serious.

The condenser is cooled by circulating cooling water through it. If an abundant supply of water is nearby, such as from a river or lake, then this can be used. Trouble may be experienced here due to water pollution. This may take the form of such things as fish or mud entering the condenser system. Filters are usually installed to cut this down, otherwise the condenser cooling water circuit may become blocked. On the other hand, the river or lake may itself become polluted by hot water returning from the condenser. If the amount of hot water is large it could have an effect on the flora and fauna of the river or lake. If a river or lake is not to hand or the risk of pollution is too high then it is common to install a *cooling tower* which is made of either wood or concrete. The hot water from the condenser is passed into the tower approximately mid-way up where it is sprayed to the bottom. Air circulates into the bottom of the tower and passes up through the water spray. Heat transfer occurs between the water and the air thus cooling the water. The warmed air passes out at the top of the tower. The cooled water is collected at the bottom of the tower from where it is pumped back to the condenser. With this method it will be noted that it is the same cooling water being circulated through the condenser. It is only necessary to make up any loss.

Looking at the steam plant system as a whole it will be noted that there are four separate circuits.

1 The furnace gas circuit

Here, air is taken into the furnace from the atmosphere in order to supply the necessary oxygen for the combustion of the boiler fuel. The combustion products pass through the boiler, transferring heat to the water in order to generate the steam. They then pass out to the atmosphere through the chimney or flue.

It should be noted that much thought and care is now being given to the control and cleanliness of the effluent from chimneys and exhaust. Gases, dust and acids formed during combustion can be carried out into the atmosphere in large quantities. These can cause undesirable effects on flora and fauna, including people, and also on buildings.

2 The steam circuit

Here, water is passed into the boiler in which it is converted into steam. It then passes into the turbine, engine or plant in which it is expanded giving up some of its energy. It is then condensed in a condenser from which it passes as condensate to be pumped back into the boiler.

3 The condenser cooling water circuit

Here, cool water passes into the condenser, has heat transferred into it by the condensing steam and then, at a higher temperature, passes out to be cooled in a river, lake, or cooling tower. Cool water is then pumped back to the condenser.

4 Cooling air circuit

Here, in the case of a cooling tower, cool air passes into the bottom of the tower from the atmosphere and heat is transferred into it from the falling hot water spray. The warm air then passes back to the atmosphere through the top of the tower. In the case of a river or lake the condenser cooling water will mix with river or lake water which will be cooled by heat transfer to the atmosphere.

It should be mentioned here that in some steam plants the condensate from the condenser is passed into a tank, called the hot well, which acts as a reservoir feed water. From the hot well, feed water is pumped through the feed pump back into the boiler. In this case, make-up water could be fed into the hot well.

5.9 The Carnot cycle and steam plant

To operate the Carnot cycle in a steam plant the processes would be as follows.

Consider the P–V diagram first, shown in Fig. 5.14(a).

A–B Water at boiler pressure P_B and volume V_A is fed from the feed pump into the boiler. This is shown as process AF. In the boiler, the water is converted into steam at pressure P_B. The volume of the steam produced is V_B. This volume of steam V_B is then fed

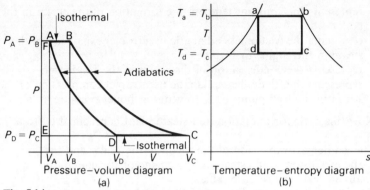

Fig. 5.14

from the boiler into the engine or turbine. This is shown as process FB.

Now the conversion of water into steam at constant pressure takes place at constant temperature, the saturation temperature T_B. This is so long as the steam does not enter the superheat phase. Hence, if the steam produced is either wet or dry saturated, then this process is isothermal.

B–C   The steam is expanded frictionless adiabatically in the engine or turbine.

C–D   The steam, after expansion, is passed from the engine or turbine into a condenser. This is shown as process CE. In the condenser the volume of the steam is reduced from V_C to V_D. This process takes place at constant condenser pressure P_C and at constant condenser saturation temperature T_C. This process is therefore isothermal.

D–A   The partially condensed steam at pressure P_C and volume V_D is fed from the condenser into the feed pump. This is shown as process ED. In the feed pump the steam is compressed frictionless adiabatically to boiler pressure P_B. This is shown as process DA. The compression converts the wet steam at condenser pressure into water at boiler pressure. This water is fed into the boiler, shown as process AF and the cycle is repeated.

Now the P–V diagram is really composed of two diagrams.

There is the engine or turbine diagram FBCE whose area will give work output. There is also the feed pump diagram EDAF whose area will give the required work input to run the feed pump.

The net work output from the plant will, therefore, be the net area of these two diagrams. This is the area ABCD.

This area ABCD is enclosed by two isothermal processes and two adiabatic processes. Hence, this is a Carnot cycle. Its thermal efficiency will be given by $(T_B - T_C)/T_B$ which is the maximum efficiency possible between these temperature limits.

The T–s diagram of the cycle is shown in Fig. 5.14(b).

a–b represents the constant temperature formation of the steam in the boiler.
b–c represents the frictionless adiabatic (isentropic) expansion of the steam in the engine or turbine.
c–d represents the condensation of the steam in the condenser.
d–a represents the frictionless adiabatic (isentropic) compression of the steam in the feed pump back to water at boiler pressure at 'a'.

Now this cycle for operation in a steam plant is practical up to a point.

The isothermal expansion of the steam in the boiler and the adiabatic expansion of the steam in the engine or turbine (more especially in turbines) is reasonable.

The impractical part is in the handling of the steam in the condenser and feed pump. In the condenser, the steam is only partially condensed and condensation must be stopped at d. Also the feed pump must be capable of handling both wet steam and water.

A slight modification to this cycle, however, will produce a cycle which is more practical although it will have a reduced thermal efficiency. This cycle is the Rankine cycle and is the usually accepted ideal cycle for steam plant.

5.10 The Rankine cycle

The modification made to the Carnot cycle to produce the Rankine cycle is that instead of stopping the condensation in the condenser at some intermediate condition, the condensation is completed. This is shown in Fig. 5.15. On the T–s diagram, Fig. 5.15(b), the Carnot cycle would be abcg. For the Rankine cycle, however, condensation is continued until it is complete at d. At this point there is all water. This

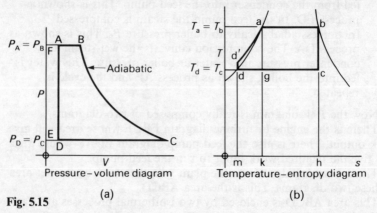

Fig. 5.15

water can be successfully dealt with in a feed pump in which its pressure can be raised in feeding it back into the boiler.

This is shown exaggerated as process d–d' on the *T–s* diagram. In the boiler the temperature of the water is raised at boiler pressure, shown as process d'–a and thus the cycle is completed. The complete Rankine cycle is, therefore, abcdd'a.

On the *P–V* diagram, Fig. 5.15(a), there are two cycles.

The work done in the engine or turbine is represented by the area FBCE. There is also the feed pump work and this is represented by the area EDAF. Now the feed pump work is negative since work must be put into the pump.

Hence,

Work done/cycle = Area ABCD   (1)

Using the steady-flow equation and neglecting changes in potential and kinetic energy, then, since for an adiabatic expansion $Q=0$, the energy equation for an adiabatic expansion becomes,

$$h_1 = h_2 + W$$

or

Specific $W = h_1 - h_2$ (2)

Using symbols as in Fig. 5.15, then,

Specific $W = h_b - h_c =$ area FBCE on *P–V* diagram   (3)

The feed pump work/unit mass = Area $EDAF = (P_B - P_C)v_D$ (4)

Hence,

Net work done/cycle $= (h_b - h_c) - (P_B - P_C)v_D$ (5)

Now the heat transfer required in the boiler to convert the water at d' into steam at b

$$= h_b - h_{d'} \quad (6)$$

But the total energy of the water entering the boiler at d'

= liquid enthalpy at d + Feed-pump work

or

$$h_{d'} = h_d + (P_B - P_C)v_D \quad (7)$$

Substituting equation (7) in equation (6)

Heat transfer required in boiler

$$= h_b - \{h_d + (P_B - P_C)v_D\}$$
$$= (h_b - h_d) - (P_B - P_C)v_D \quad (8)$$

Now, thermal efficiency of cycle

$$= \frac{\text{net work done/cycle}}{\text{heat received/cycle}}$$

Hence, from equations (5) and (8),

Thermal efficiency of Rankine cycle

$$= \frac{(h_b - h_c) - (P_B - P_C)v_D}{(h_b - h_d) - (P_B - P_C)v_D} \qquad (9)$$

The feed pump term $(P_B - P_C)v_D$, is, however, small compared with the other energy quantities and hence it can be sensibly neglected.

Thus, equation (9) becomes,

Thermal, or now, the Rankine efficiency

$$= \frac{(h_b - h_c)}{(h_b - h_d)} \qquad (10)$$

The cycle is named after William John Rankine (1820–72), a Glasgow University Professor.

If the work done by the feed pump is neglected and assuming that the steam expansion can be expressed in the form $PV^n = $ constant, then the

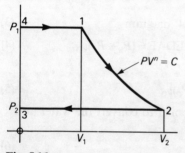

Fig. 5.16

P–V diagram for the Rankine cycle is as shown in Fig. 5.16. From this diagram,

Net work done/cycle = Area under 4–1 + area under 1–2 –
 – area under 2–3

$$= P_1 V_1 + \frac{P_1 V_1 - P_2 V_2}{n-1} - P_2 V_2$$

$$= (P_1 V_1 - P_2 V_2) + \frac{P_1 V_1 - P_2 V_2}{n-1}$$

$$= (P_1 V_1 - P_2 V_2)\left(1 + \frac{1}{n-1}\right)$$

$$= (P_1V_1 - P_2V_2)\frac{(n-1)+1}{n-1}$$

$$= \frac{n}{n-1}(P_1V_1 - P_2V_2) \qquad (11)$$

If the superheated steam is used in the Rankine cycle then the appearance of the cycle on the T–s diagram is shown in Fig. 5.17. The difference between this diagram and the previous one is the inclusion of the superheat line bc. The complete cycle is now abcdee'.

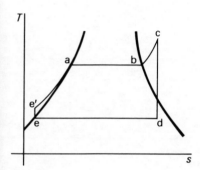

Fig. 5.17

The thermal efficiency has the same form as before. Using the lettering of Fig. 5.17.

Rankine $\eta = \dfrac{h_c - h_d}{h_c - h_e}$ \qquad (12)

There is little gain in thermal efficiency as a result of using superheated steam over that of using saturated steam. The chief advantage of using superheated steam are, firstly, there is little or no condensation loss in transmission. Secondly, there is greater potential for enthalpy drop and hence work done.

Note, also, that by using superheated steam, there is a further departure from the Carnot cycle since the final temperature of the steam is above the constant saturation temperature of the boiler.

Note, further, that the Rankine cycle will have a high work ratio, ($\rightarrow$ unity), since the net work done/cycle is very close to the positive work done/cycle, the feed pump work being very low by comparison. Also, the Rankine cycle will have a higher work ratio than the Carnot vapour cycle.

Example 26

A steam turbine in a plant operating on the Rankine cycle receives steam from the boiler at a pressure of 4 MN/m² and at a temperature

of 400 °C. The exhaust to the condenser is at a pressure of 50 kN/m². From the condenser the condensate is returned to the boiler. Neglecting losses, determine:
(a) the energy supplied in the boiler/kg of steam;
(b) the dryness fraction of the steam as it leaves the turbine to enter the condenser;
(c) the specific net work output;
(d) the Rankine efficiency;
(e) the specific steam consumption.

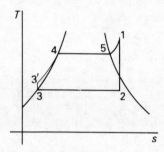

(a) Energy supplied by boiler $= (h_1 - h_3)$ kJ/kg
From steam tables,
At 4 MN/m², $h_1 = \underline{3\,214\text{ kJ/kg}}$
At 50 kN/m², $h_3 = \underline{340\text{ kJ/kg}}$

$\therefore$ Energy supplied by boiler $= 3\,214 - 340$

$= \underline{2\,074\text{ kJ/kg}}$

(b) The expansion through the turbine is theoretically isentropic.
$\therefore s_1 = s_2$
and,
$s_1 = 6.769$ kJ/kgK
$s_2 = s_{f2} + (x_2 \times s_{fg2})$ where $x_2 =$ dryness fraction
$= 1.091 + 6.502 x_2$
$\therefore 1.091 + 6.502 x_2 = 6.769$
hence,
$$x_2 = \frac{6.769 - 1.091}{6.502} = \frac{5.678}{6.502} = \underline{0.87}$$

(c) Specific net work output $= h_1 - h_2$ and,
$h_2 = h_{f2} + x_2 h_{fg2}$

$$= 340 + (0.87 \times 2\,305)$$
$$= 340 + 2\,005.4$$
$$= 2\,345.4 \text{ kJ/kg}$$

∴ Specific net work output $= 3\,214 - 2\,345.4$
$$= \underline{868.6 \text{ kJ/kg}} = W_s$$

(d) Rankine $\eta = \dfrac{h_1 - h_2}{h_1 - h_3}$

$$= \dfrac{868.6}{3\,214 - 340} = \dfrac{868.6}{2\,874}$$

$$= \underline{0.302}$$

$$= 0.302 \times 100\% = \underline{30.2\%}$$

(e) Specific steam consumption $= \dfrac{3\,600}{W_s}$ (see section 5.1, equation (9))

$$= \dfrac{3\,600}{868.6}$$

$$= \underline{4.15 \text{ kg/kWh}}$$

Example 27

A steam turbine plant operates on the ideal Rankine cycle and has a power output of 250 MW. The steam enters the turbine at a pressure of 3 MN/m² and with a temperature of 350 °C. The steam leaves the turbine and enters the condenser at a pressure of 20 kN/m². For this plant, determine:
(a) the dryness fraction of the steam as it leaves the turbine;
(b) the Rankine efficiency;
(c) the specific steam consumption;
(d) the mass of steam used/hour.

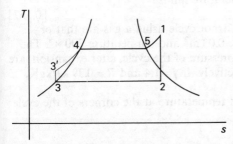

(a) $h_1 = 3\,117\text{ kJ/kg}$

$\quad = s_1 = s_2$

$\therefore\ 6.744 = 0.832 + (x_2 \times 7.075)$

hence,

$$x_2 = \frac{6.744 - 0.832}{7.075} = \frac{5.912}{7.075} = \underline{0.84}$$

= dryness fraction leaving turbine.

(b) $h_2 = 251 + (0.84 \times 2\,358) = 251 + 1\,981 = \underline{2\,232\text{ kJ/kg}}$

$h_3 = \underline{251\text{ kJ/kg}}$

Rankine $\eta = \dfrac{h_1 - h_2}{h_1 - h_3} = \dfrac{3\,117 - 2\,232}{3\,117 - 251} = \dfrac{885}{2\,866}$

$\quad = \underline{0.308}$

$\quad = 0.308 \times 100\% = \underline{30.8\%}$

(c) Specific net work done $= h_1 - h_2 = \underline{885\text{ kJ/kg}}$

Specific steam consumption $= \dfrac{3\,600}{885}$ (see section 5.1, equation (9))

$\quad = \underline{4.068\text{ kg/kWh}}$

(d) $\therefore$ mass of steam used/h $= 4\,068 \times 250 \times 10^3 = \underline{1\,017\,000\text{ kg}}$

$\qquad\qquad\qquad\qquad\qquad\qquad\qquad = \underline{1\,017\text{ tonne}}$

Questions 5

1 A Carnot cycle has the temperature limits of 400 °C and 25 °C. Determine the thermal efficiency of the cycle.

[0.557]

2 A cycle has a maximum temperature of 450 °C and a minimum temperature of 40 °C. What is the maximum thermal efficiency possible between these temperature limits?

[0.567]

3 The initial conditions of a Carnot cycle using a gas are that of pressure 1.4 MN/m^2, volume 0.03 m^3 and temperature 300 °C. The minimum temperature and pressure of the cycle, after expansion, are 50 °C and 150 kN/m^2, respectively. If $\gamma = 1.4$ and $R = 0.29$ kJ/kgK, determine:
(a) the pressure, volume and temperature at the corners of the cycle;
(b) the mass of gas;
(c) the thermal efficiency;

(d) the net work done/cycle;
(e) the work ratio.

[(a) $P_1 = 1.4$ MN/m² $\quad V_1 = 0.03$ m³ $\quad t_1 = 300$ °C
$P_2 = 1.117$ MN/m² $\quad V_2 = 0.0376$ m³ $\quad t_2 = 300$ °C
$P_3 = 0.15$ MN/m² $\quad V_3 = 0.1578$ m³ $\quad t_3 = 50$ °C
$P_4 = 0.188$ MN/m² $\quad V_4 = 0.126$ m³ $\quad t_4 = 50$ °C
(b) 0.253 kg, (c) 0.436, (d) 4.145 kJ, (e) 0.075]

4 An engine working on a constant volume cycle has a relative efficiency of 49% and an actual thermal efficiency of 28%. Determine:
(a) the theoretical thermal efficiency of the engine cycle;
(b) the compression ratio of the theoretical engine cycle.
Take $\gamma = 1.4$
[(a) 57.1% (b) 8.3:1]

5 0.05 kg of air is taken through an ideal constant volume cycle. Conditions at the beginning of the adiabatic compression are that of pressure = 130 kN/m² and temperature = 25 °C. The pressure at the end of the initial compression is 2 MN/m². The maximum pressure of the cycle is 4.2 MN/m². For this cycle, determine:
(a) the net work done;
(b) the thermal efficiency;
(c) the work ratio;
(d) the mean effective pressure.
Take $c_p = 1.005$ kJ/kgK, $c_v = 0.716$ kJ/kgK.

[(a) 14.01 kJ, (b) 54.4%, (c) 0.525, (d) 495.05 kN/m²]

6 An ideal constant pressure cycle, using air, has a maximum volume of 0.02 m³. The pressure volume and temperature of the air at the beginning of the adiabatic compression are 100 kN/m², 0.01 m³ and 25 °C, respectively. The pressure ratio of the cycle is 8:1. For the air, take $c_p = 1.005$ kJ/kgK, $c_v = 0.714$ kJ/kgK. Determine:
(a) the pressure, volume and temperature at the cycle change points;
(b) the net work done/cycle;
(c) the thermal efficiency of the cycle;
(d) the work ratio of the cycle;
(e) the mean effective pressure of the cycle.

[(a) $P_1 = 100$ kN/m² $\quad V_1 = 0.01$ m³ $\quad t_1 = 25$ °C
$P_2 = 800$ kN/m² $\quad V_2 = 0.0023$ m³ $\quad t_2 = 271.45$ °C
$P_3 = 800$ kN/m² $\quad V_3 = 0.0046$ m³ $\quad t_3 = 815.9$ °C
$P_4 = 100$ kN/m² $\quad V_4 = 0.02$ m³ $\quad t_4 = 323$ °C
(b) 2.848 kJ, (c) 45.3%, (d) 0.478, (e) 160.9 kN/m²]

7 An open circuit, continuous combustion, constant pressure, gas turbine, using air, has an intake pressure and temperature of

100 kN/m² and 15 °C, respectively. The air is compressed adiabatically in a rotary compressor to a pressure of 700 kN/m² with an isentropic efficiency of 80%. From the compressor the air passes through a combustion chamber at constant pressure in which the temperature is raised to 900 °C. The air then passes through a gas turbine in which it is expanded adiabatically to an exhaust pressure of 100 kN/m² with an isentropic efficiency of 83%. A net output of 370 kW is required from the plant. Neglect the mass of fuel.

Take, $c_p = 1.004$ kJ/kgK, $c_v = 0.709$ kJ/kgK, throughout. Determine:
(a) the required mass flow rate of the air in kg/s;
(b) the thermal efficiency;
(c) the work ratio.
[(a) 2.52 kg/s, (b) 24.1%, (c) 0.345]

8 A gas turbine operating on the ideal constant pressure cycle has a pressure ratio of 6:1 which gives the maximum net work output within the temperature limits of the cycle. The lower temperature limit is 18 °C. Compression in the compressor and expansion in the turbine are isentropic.

Take, $\gamma = 1.39$, $c_p = 1.004$ kJ/kgK. Determine:
(a) the maximum temperature of the cycle;
(b) the maximum net specific work output;
(c) the thermal efficiency and work ratio at maximum specific work output.
[(a) 522.3 °C, (b) 124.8 kJ/kg. (c) 39.5%, 0.395]

9 A gas turbine operating on an ideal constant pressure cycle has a pressure ratio of 6:1. The relative effieicncy of the plant is 60% and the specific fuel consumption is 0.32 kg/kWh. The fuel used has a calorific value of 42.5 MJ/kg. If $\gamma = 1.39$, determine the actual thermal efficiency of the plant.
[26.5%]

10 An engine works on the ideal Diesel cycle. The overall volume ratio of the cycle is 14:1. Constant pressure heat addition is 600 kJ/cycle and it ceases at 0.2 of the stroke. Take $\gamma = 1.4$. Determine:
(a) the thermal efficiency of the cycle;
(b) the heat rejected/cycle;
(c) the net work output/cycle.
[(a) 52%, (b) 288 kJ, (c) 312 kJ]

11 0.25 kg of air is taken through an ideal Diesel cycle. The pressure and temperature at the commencement of the adiabatic compression are 130 kN/m² and 35 °C, respectively. The pressure at the end of the adiabatic compression is 3.5 MN/m². The pressure at the end of the adiabatic expansion is 380 kN/m².

Take, $c_p = 1.005$ kJ/kgK, $c_v = 0.715$ kJ/kgK. Determine:
(a) the pressure, volume and temperature at the cycle change points;

(b) the net work done/cycle;
(c) the thermal efficiency of the cycle;
(d) the work ratio of the cycle;
(e) the mean effective pressure of the cycle.

[(a) $P_1 = 130 \text{ kN/m}^2$ $V_1 = 0.172 \text{ m}^3$ $t_1 = 35\,°\text{C}$
$P_2 = 3.5 \text{ MN/m}^2$ $V_2 = 0.016\,5 \text{ m}^3$ $t_2 = 524.7\,°\text{C}$
$P_3 = 3.5 \text{ MN/m}^2$ $V_3 = 0.035 \text{ m}^3$ $t_3 = 1\,446.6\,°\text{C}$
$P_4 = 380 \text{ kN/m}^2$ $V_4 = 0.172 \text{ m}^3$ $t_4 = 627.3\,°\text{C}$
(b) 125.75 kJ, (c) 54.3%, (d) 0.612, (e) 808.7 kN/m^2]

12 The pressure, volume and temperature at the beginning of the adiabatic compression of an ideal dual combustion cycle are 105 kN/m^2, 0.25 m^3 and 25 °C, respectively. At the end of the adiabatic compression the volume is 0.02 m^3. The pressure ratio of the constant volume heat addition process is 1.6:1. The volume ratio of the constant pressure heat addition process is 1.7:1. Take $\gamma = 1.39$ and $R = 0.28$ kJ/kgK. Determine:
(a) the pressure, volume and temperature at the cycle change points;
(b) the net work done/cycle;
(c) the thermal efficiency of the cycle;
(d) the work ratio of the cycle;
(e) the mean effective pressure of the cycle;
(f) the maximum thermal efficiency possible within the temperature limits of the cycle.

[(a) $P_1 = 105 \text{ kN/m}^2$ $V_1 = 0.25 \text{ m}^3$ $t_1 = 25\,°\text{C}$
$P_2 = 3.514 \text{ MN/m}^2$ $V_2 = 0.02 \text{ m}^3$ $t_2 = 525.6\,°\text{C}$
$P_3 = 5.62 \text{ MN/m}^2$ $V_3 = 0.02 \text{ m}^3$ $t_3 = 1\,004.8\,°\text{C}$
$P_4 = 5.62 \text{ MN/m}^2$ $V_4 = 0.034 \text{ m}^3$ $t_4 = 1\,899\,°\text{C}$
$P_5 = 350 \text{ kN/m}^2$ $V_5 = 0.25 \text{ m}^3$ $t_5 = 723\,°\text{C}$
(b) 231.6 kJ, (c) 59.5%, (d) 0.673, (e) 1.007 MN/m^2, (f) 86.3%]

13 An ideal Stirling cycle, using regeneration, uses 0.7 kg of hydrogen. The initial pressure and volume of the cycle are 4.5 MN/m^2 and 0.25 m^3, respectively. The maximum temperature of the cycle is 650 °C. The minimum volume of the cycle is 0.05 m^3. Take $R = 4.12$ kJ/khK. Determine:
(a) the pressure, volume and temperature at the cycle change points;
(b) the net work done/cycle;
(c) the ideal thermal efficiency of the cycle.

[(a) $P_1 = 4.5 \text{ MN/m}^2$ $V_1 = 0.25 \text{ m}^3$ $t_1 = 117.1\,°\text{C}$
$P_2 = 22.5 \text{ MN/m}^2$ $V_2 = 0.05 \text{ m}^3$ $t_2 = 117.1\,°\text{C}$
$P_3 = 53.2 \text{ MN/m}^2$ $V_3 = 0.05 \text{ m}^3$ $t_3 = 650\,°\text{C}$
$P_4 = 10.64 \text{ MN/m}^2$ $V_4 = 0.25 \text{ m}^3$ $t_4 = 650\,°\text{C}$
(b) 2.474 MJ, (c) 57.7%]

14 A steam turbine plant operates on the ideal Rankine cycle and uses 800 tonne of steam/hour. The steam enters the turbine at a pressure

of 3 MN/m² and with a temperature of 400 °C. The steam leaves the turbine and enters the condenser at a pressure of 30 kN/m². For this plant, determine:
(a) the dryness fraction of the steam as it leaves the turbine;
(b) the Rankine efficiency;
(c) the specific steam consumption;
(d) the theoretical power output of the turbine.

[(a) 0.88, (b) 30.1%, (c) 4.06 kg/kWh, (d) 197 MW]

Chapter 6

Refrigerators and heat pumps

6.1 The refrigeration process

If a body is to be maintained at a temperature lower than that of its surrounding, or ambient temperature, then, any heat transfer which will naturally occur down the temperature gradient from the surroundings to the body (Second Law of Thermodynamics), must be transferred back to the surroundings. Unless this is done, the temperature of the body will increase to that of its surroundings.

Now the transfer of heat from a colder to a hotter body is contrary to the Second Law of Thermodynamics, which implies that, if such a transfer of heat is required, then, external energy is required to facilitate the transfer. This external energy can be supplied either by means of a heating device as a compressor (or pump), the use of either producing the necessary increase in temperature.

The cyclic process by which natural heat transfer down a temperature gradient is returned up the temperature gradient by means of the supply of external energy is the process of *refrigeration*.

In any *refrigerator*, as the plant is called, there will be an amount of energy removed from the cold body by the refrigeration process. This is referred to as the *refrigeration effect*.

The ratio: $\dfrac{\text{Refrigerating effect}}{\text{External energy supplied}}$

is called the *coefficient of performance* (*C.O.P.*).

It will be noted that this definition is similar to that used for efficiency. The term efficiency is not used here, however, because very often C.O.P. > 1, and hence the term coefficient is used rather than efficiency.

The various heat transfers associated with the refrigeration process are illustrated in Fig. 6.1. Note that the high temperature is higher than the ambient temperature in order that heat transfer can take place.

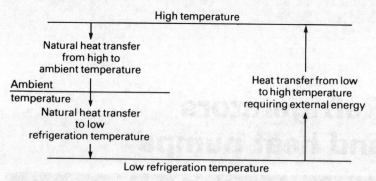

Fig. 6.1

The heat transfer from the high to the low refrigeration temperature takes place in two stages. There is a natural heat transfer to the surroundings from the high to ambient temperature. This is followed by a natural heat transfer from ambient to the low refrigeration temperature. The heat transfer from the low to high temperature, requiring external energy, is direct.

The working substance which flows through a refrigerator is called a *refrigerant*. It is usual that heat transfer into the refrigerant at low temperature evaporates the refrigerant. Heat transfer from the refrigerant at high temperature condenses the refrigerant.

Commonly used refrigerants are Ammonia (NH_3), Methyl Chloride (CH_3Cl), Freon-12 (Dichlorodifluoromethane, CCl_2F_2) and Carbon Dioxide (CO_2). These substances remain as liquids and can be evaporated at suitable sub-zero low temperatures which make them suitable for use as refrigerants.

When working at very low temperatures, the term *cryogenics* is used rather than *refrigeration*.

It will be noted that the refrigeration cycle is the reverse of the heat engine cycle. In the heat engine cycle energy is received at high temperature, rejected at low temperature and work is obtained from the cycle. In the refrigeration cycle, however, energy is received at low temperature, rejected at high temperature and work (or heat) is required to perform the cycle. Due to the transfer of energy from low to high temperature, the refrigerator is sometimes referred to as a *heat pump*.

6.2 The reversed Carnot cycle

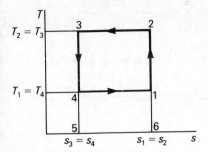

Fig. 6.2

In Fig. 6.2 is illustrated a T–s diagram of a reversed Carnot cycle. The processes of the cycle are as follows:

4–1 Isothermal expansion at low temperature $T_1 = T_4$

For an isothermal process, $Q = W$
$$\therefore\ T_1(s_1 - s_4) = Q_{4-1} = W_{4-1} = \text{area } 5416$$

1–2 Isentropic compression from T_1 to T_2.

The compression is also adiabatic.
$$\therefore\ Q_{1-2} = 0 \quad \text{and} \quad W_{1-2} = -U_{1-2}$$

2–3 Isothermal compression at high temperature $T_2 = T_3$

Also, $-T_2(s_2 - s_3) = -T_2(s_1 - s_4) = -Q_{2-3} = -W_{2-3} = \text{area } 5623$

(−ve sign because heat transfer is negative, i.e. heat is lost).

3–4 Isentropic expansion from T_3 to T_4 (same as T_2 to T_1).

The expansion is also adiabatic.
$$\therefore\ Q_{3-4} = 0 \quad \text{and} \quad W_{3-4} = -U_{3-4}$$

For this cycle,

Heat received at low temperature = Refrigerating effect = $T_1(s_1 - s_4)$.

Now, for a cycle,

$$\oint W = \oint Q$$

or Net work = Heat received − Heat rejected

In this case,

$$\text{Net work} = \oint W = T_1(s_1 - s_4) - T_2(s_1 - s_4)$$
$$= -(T_2 - T_1)(s_1 - s_4)$$

The $-$ve sign shows that work must be supplied in order to perform the cycle.

Thus, the external energy supplied to perform the cycle
$$= (T_2 - T_1)(s_1 - s_4)$$
For a refrigeration cycle,

$$\text{C.O.P.} = \frac{\text{Refrigerating effect}}{\text{External energy supplied}}$$

Hence, in this case,

$$\text{C.O.P.} = \frac{T_1(s_1 - s_4)}{(T_2 - T_1)(s_1 - s_4)}$$
$$= \frac{T_1}{T_2 - T_1}$$

Now, the Carnot cycle is composed of reversible processes which are the most efficient thermodynamic processes possible (see Chapter 3).

Hence, the reversed Carnot cycle will have the highest C.O.P. possible between any given limits of temperature.

Note that the equation,

$$\text{C.O.P.} = \frac{T_1}{T_2 - T_1}$$

can be written,

$$\text{C.O.P.} = \frac{1}{\dfrac{T_2}{T_1} - 1}$$

This shows that as $T_1 \to T_2$, so C.O.P. $\to \infty$.

Thus, to improve the C.O.P. of a refrigerator, the limits of temperature must be as close as possible or, in other words, do not refrigerate at a lower temperature than is necessary.

Further, since in this case, as $T_1 \to T_2$, so C.O.P. $\to \infty$, it shows that it is theoretically possible to have C.O.P. values > 1.

6.3 Reversed constant pressure cycle

Some early attempts to carry out the refrigeration process were made using reversed air engines.

One successful refrigerator resulted from the use of the reversed constant pressure (Joule) cycle. This was the Bell–Coleman refrigerator of about 1880 (see section 5.4).

The P–V and T–s diagrams for the reversed constant pressure cycle are illustrated in Fig. 6.3.

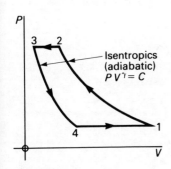
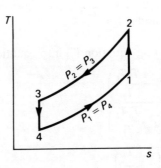

Fig. 6.3

For the adiabatic process of this cycle, $Q = 0$. The constant pressure process, 4–1, produces the refrigerating effect.

Refrigerating effect $= \dot{m}c_p(T_1 - T_4)$ (1)

Energy rejected as heat transfer during process, 2–3 $= \dot{m}c_p(T_2 - T_3)$ (2)

External energy supplied = Energy rejected − Refrigerating effect

$$= \dot{m}c_p(T_2 - T_3) - \dot{m}c_p(T_1 - T_4)$$
$$= \dot{m}c_p[(T_2 - T_3) - (T_1 - T_4)] \quad (3)$$

assuming c_p constant.

From this,

$$\text{C.O.P.} = \frac{\dot{m}c_p(T_1 - T_4)}{\dot{m}c_p[(T_2 - T_3) - (T_1 - T_4)]}$$

$$= \frac{T_1 - T_4}{(T_2 - T_3) - (T_1 - T_4)} \quad (4)$$

Now for the adiabatic processes,

$$\frac{T_3}{T_4} = \left(\frac{P_3}{P_4}\right)^{\gamma - 1/\gamma} \quad \text{and} \quad \frac{T_2}{T_1} = \left(\frac{P_2}{P_1}\right)^{\gamma - 1/\gamma}$$

But, $P_3 = P_2$ and $P_4 = P_1$

$$\therefore \frac{T_3}{T_4} = \frac{T_2}{T_1} \tag{5}$$

from equation (4)

$$\text{C.O.P.} = \frac{(T_1 - T_4)}{(T_2 - T_3)} - 1 \tag{6}$$

from equation (5)

$$T_4 = \frac{T_3 T_1}{T_2} \quad \text{and} \quad T_3 = \frac{T_4 T_2}{T_1} \tag{7}$$

Substituting equation (7) into equation (6)

$$\text{C.O.P.} = \frac{\left(T_1 - \dfrac{T_3 T_1}{T_2}\right)}{\left(T_2 - \dfrac{T_4 T_2}{T_1}\right)} - 1$$

$$= \frac{T_1\left(1 - \dfrac{T_3}{T_2}\right)}{T_2\left(1 - \dfrac{T_4}{T_1}\right)} - 1 \tag{8}$$

But from equations (5)

$$\frac{T_3}{T_2} = \frac{T_4}{T_1}$$

and hence, equation (8) becomes,

$$\text{C.O.P.} = \frac{T_1}{T_2} - 1$$

$$= \frac{T_1}{T_2 - T_1} \tag{9}$$

This C.O.P. is less than that of the reversed Carnot cycle within the same temperature limits.

In this case,

$$\text{C.O.P. (Carnot)} = \frac{T_4}{T_2 - T_4}$$

and processes 4–1 and 2–3 would be isothermal.

Air is not a commonly used refrigerant. It has the disadvantage of a moisture content. This moisture will freeze at 0 °C and could eventually block the refrigerator pipe-work and valves.

Air drying equipment can be installed, but it is doubtful whether the moisture can be totally removed. Also, air has poor heat transfer properties and in a refrigeration plant, good heat transfer is required.

It should be noted, however, that in air conditioning plant in which cooled air is required, such a circuit as indicated can be used. Turbo-type machinery would probably be used instead of reciprocating machinery in many cases. Cooled air in this case is passed directly into the air conditioned chamber.

An arrangement of this type can often be found in modern aircraft where air conditioning of the cockpit and cabin is required.

6.4 The vapour compression refrigerator

In the vapour-compression refrigerator, as the name implies, liquid refrigerants are used which are alternately evaporated and condensed.

Using a liquid refrigerant, the Carnot cycle could be closely approximated. This is illustrated in Fig. 6.4. It has been shown in work

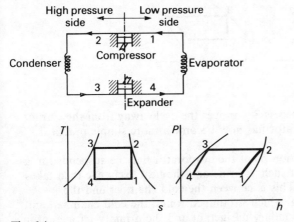

Fig. 6.4

on two-phase systems that during the evaporation of a liquid at constant pressure the temperature remains constant. Referring to Fig. 6.4, then, a wet low-pressure, low-temperature refrigerant enters the evaporator at 4 in which it is evaporated to a nearly dry state at 1. This evaporation process produces the refrigerating effect. The refrigerant then enters a compressor in which it is compressed, theoretically isentropically, to 2. As illustrated, the refrigerant would then be dry saturated at a higher pressure and temperature. The refrigerant then passes through a condenser at constant pressure and temperature and is condensed to liquid at 3. The refrigerant then passes through an expander in which it is expanded, theoretically isentropically, back to its original low-pressure, low-temperature, wet state at 4.

Temperature–entropy (T–s) and pressure–enthalpy (P–h) diagrams of the cycle are also illustrated in Fig. 6.4.

It is common practice, however, to use a throttle valve or regulator in place of the expander. This is illustrated in Fig. 6.5 and most vapour compression refrigerators have this basic arrangement.

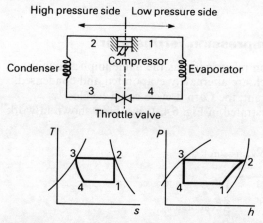

Fig. 6.5

The throttling process 3–4 moves the cycle away from the Carnot cycle but the refrigerator has now become a more simple practical arrangement.

In large refrigeration plant the evaporator may be suspended in a secondary refrigerant such as brine and the heat exchange then takes place in two stages. This is between the cold chamber and the secondary refrigerant, which is pumped round the cold chamber, and then between the secondary refrigerant and the primary refrigerant in the evaporator in the refrigerator. Again, in large refrigeration plant, the condenser may be water cooled or have forced-draught air cooling using fans.

In small refrigeration plant, such as in the domestic refrigerator, the evaporator is suspended directly in the cold chamber and the condenser is suspended in the surrounding atmospheric air. Also, in small refrigeration plant, the throttling process may be accomplished by using a short length of capillary tubing. This produces a fixed low temperature in the evaporator. The control of the cold chamber temperature is obtained by using a thermostat in the cold chamber. When the required temperature is reached in the cold chamber, controls connected to the thermostat, switch off the motor driving the refrigerator. The temperature in the cold chamber then slowly rises and the thermostat controls then switch on the motor and the process is then repeated. If a throttle valve is fitted, then there is a control on the evaporator temperature.

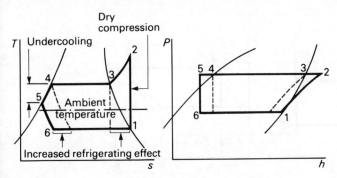

Fig. 6.6

In Fig. 6.6 is shown the T–s and P–h diagrams of the type of cycle more commonly used in the vapour compression refrigerator. The modifications made to the cycle already illustrated in Fig. 6.5 produce a more effective operation of the plant. It will be noted that at entry to the compressor at 1, the refrigerant is shown as being dry saturated. Sometimes there is a slight degree of superheat. The effect of this is to increase the refrigerating effect and also to produce dry compression in the refrigerator, shown as process 1–2. This means that there is no loss of mass flow due to evaporation of the liquid refrigerant in the compressor during the induction stroke. Also liquid refrigerant can wash lubricant from the cylinder walls and also carry over some lubricant into the other sections of the plant which results in a reduction of heat transfer. A further improvement can be obtained by undercooling (or subcooling), the refrigerant after condensation, shown as process 4–5. Here, the refrigerant is cooled toward the ambient temperature. This produces a wetter vapour, at 6, after the throttling process and an improved refrigerating effect follows.

It should be noted that the refrigerating effect per unit time is called the *duty* of the refrigerator. This will depend upon the end states of the refrigerant in the evaporator and also the mass flow rate of the refrigerant.

Tables of properties for various refrigerants exist and are of a similar nature to the tables for the properties of steam (or water substance, as it is sometimes called). The refrigerant tables have their own reference state which is commonly that both the specific enthalpy and specific entropy are considered as being zero at $-40\,°C$.

It should be further noted that some refrigeration plants have a more complex circuit arrangement than the simple circuit illustrated.

6.5 Calculations for the vapour compression refrigerator

The cycle illustrated in Fig. 6.7 is representative of a typical vapour compression cycle. An analysis of this cycle follows.

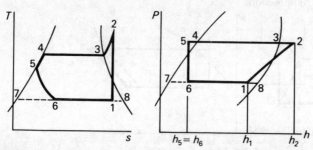

Fig. 6.7

As already stated, tables of the properties of refrigerants are available.

Hence, the properties of state points 7, 8, 4 and 3 may be looked up in the relevant tables.

At state point 1.

This is at exit from the evaporator and entry to the compressor.

$h_1 = h_{f7} + x_1(h_{g8} - h_{f7})$

$s_1 = s_{f7} + x_1(s_{g8} - s_{f7})$

The compression, 1–2, is considered as being theoretically isentropic, hence,

$s_1 = s_2$

The specific volume of the refrigerant at entry to the compressor at 1, together with the compressor characteristics, will control the mass flow of refrigerant through the refrigerator.

$v_1 = x_1 v_{g8}$

At state point 2.

This is at delivery from the compressor and entry to the condenser.

h_2 may be determined from superheat tables.

Alternatively,

If c_{pv} = specific heat capacity of the superheated vapour,

$h_2 = h_{g3} + c_{pv}(T_2 - T_3)$

Since the compression is isentropic

$s_2 = s_1$

s_2 may be determined from superheat tables.

Alternatively,

$$s_2 = s_{g3} + c_{pv} \ln \frac{T_2}{T_3}$$

At state point 5.
This is at exit from the condenser and entry to the throttle valve.

If c_{pl} = specific heat capacity of the liquid,

$h_5 = h_{f4} - c_{pl}(T_4 - T_5)$

Alternatively, h_5 may be looked up in the tables as the specific enthalpy of the liquid refrigerant at saturation temperature T_5.

Now the process, 5–6, is a throttling process. Hence,

$h_5 = h_6$

At state point 6.
This is at exit from the throttle valve and entry to the evaporator.

Because of the throttling process,

$h_6 = h_5$ (see *Thermodynamics, Level 3*, section 3.24)

Alternatively,

$h_6 = h_{f7} + x_6(h_{g8} - h_{f7})$

Alternatively,

$h_6 = h_1 - (h_1 - h_6)$

$= h_1 -$ specific refrigerating effect.

From the information obtained,

Theoretical C.O.P. $= \dfrac{h_1 - h_6}{h_2 - h_1}$

In a refrigerator trial,

Actual C.O.P. $= \dfrac{\text{Actual refrigerating effect}}{\text{Actual energy input}}$

6.6 The heat pump

During the analysis of the refrigeration process it will be noted that more energy is rejected at the high temperature than is required to drive the refrigerator.

If the temperature during the rejection process is sufficiently high it can be considered that the heat transfer during this process could be usefully used in a warming process. This heat transfer being greater than the energy required to drive the plant presents an attractive idea.

This concept was suggested by Lord Kelvin in 1852.

The vapour compression refrigerator with suitably arranged pressures and temperatures can be considered as being suitable for a *heat pump*. Many commercial machines have been manufactured using this process. The evaporator in this case is buried under the soil or suspended in a river, a lake or atmospheric air.

The heat pump has not gained a wide acceptance. It is more complex and more difficult to run and maintain than some of the more conventional heating systems.

However, a decrease in fossil fuel availability could encourage the further development and use of this apparatus.

Example 28

A vapour compression refrigerator circulates 0.1 kg of ammonia/s. Condensation takes place at 30 °C and evaporation at −15 °C. There is no undercooling after condensation. The temperature after compression is 70 °C and the specific heat capacity of the superheated vapour is 2.82 kJ/kgK. Determine:
(a) the coefficient of performance;
(b) the ice produced in kg/h from water at 18 °C to ice at 0 °C. Specific heat capacity of water = 4.187 kJ/kg. Specific enthalpy of fusion of ice = 336 kJ/kg;
(c) the effective swept volume of the compressor in m³/min.
The relevant properties of ammonia are as follows:

Sat. temp. °C	Spec. enthalpy kJ/kg		Spec. entropy kJ/kgK		Spec. Vol. m³/kg	
	h_f	h_g	s_f	s_g	v_f	v_g
−15	112.3	1 426	0.457	5.549	0.001 52	0.509
30	323.1	1 469	1.204	4.984	0.001 68	0.111

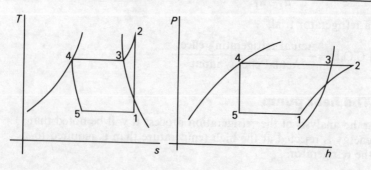

(a) $h_2 = 1\,469 + 2.82(70 - 30) = 1\,469 + (2.82 \times 40)$

$\quad = 1\,469 + 112.8 = \underline{1\,581.8 \text{ kJ/kg}}$

$s_2 = 4.984 + 2.82 \ln \dfrac{343}{303} \qquad T_3 = 30 + 273 = \underline{303 \text{ K}}$

$\quad = 4.984 + 2.82 \ln 1.132 \qquad T_2 = 70 + 273 = \underline{343 \text{ K}}$

$\quad = 4.984 + (2.82 \times 0.124) = 4.984 + 0.349$

$\quad = \underline{5.333 \text{ kJ/kgK}}$

$s_2 = s_1$

$\therefore\; 5.333 = 0.457 + x_1(5.549 - 0.457)$

$\quad x_1 = \dfrac{5.333 - 0.457}{5.549 - 0.457} = \dfrac{4.876}{5.092}$

$\quad = \underline{0.957}$

$h_1 = 112.3 + 0.957(1\,426 - 112.3)$

$\quad = 112.3 + (0.957 \times 1\,313.7) = 112.3 + 1\,257.2$

$\quad = \underline{1\,369.5 \text{ kJ/kg}}$

$h_4 = h_5 = \underline{323.1 \text{ kJ/kg}}$

C.O.P. $= \dfrac{h_1 - h_5}{h_2 - h_1} = \dfrac{1\,369.5 - 323.1}{1\,581.8 - 1\,369.5} = \dfrac{1\,046.4}{212.3}$

$\quad = \underline{4.93}$

(b) Refrigerating effect/h $= 0.1 \times 3\,600 \times (h_1 - h_5)$

$\quad\quad\quad\quad\quad\quad\quad\quad = \underline{(0.1 \times 3\,600 \times 1\,046.4) \text{ kJ}}$

Refrigerating effect required/kg ice

$\quad = (18 \times 4.187) + 336 = 75.37 + 336$

$\quad = \underline{411.37 \text{ kJ}}$

$\therefore\;$ ice produced $= \dfrac{0.1 \times 3\,600 \times 1\,046.4}{411.37}$

$\quad\quad\quad\quad\quad\quad = \underline{915.7 \text{ kg/h}}$

(c) Specific volume of refrigerant entering compressor

$\quad = v_1 = 0.957 \times 0.509 = \underline{0.487 \text{ m}^3/\text{kg}}$

$\therefore\;$ Effective swept volume of compressor

$\quad = 0.1 \times 60 \times 0.487$

$\quad = \underline{2.92 \text{ m}^3/\text{min}}$

Example 29

A heat pump consists of an evaporator, compressor, condenser and throttle regulator and circulates Freon 12. The pressure limits of the heat pump are 0.362 MN/m² and 0.745 MN/m². The heating effect required from the condenser unit is 50 MJ/h. The Freon 12 is assumed dry saturated at the beginning of compression. At the end of the condensation process the refrigerant is liquid but not undercooled. The specific heat capacity of the superheated vapour can be assumed constant at 0.71 kJ/kgK. Determine:
(a) the mass flow of Freon 12 in kg/h;
(b) the dryness fraction of the Freon 12 as it enters the evaporator;
(c) the power required from the driving motor assuming that only 75% of the power of the driving motor appears in the Freon 12;
(d) the cost of running the heat pump/eight hour day if the driving motor is driven electrically and the electricity cost is 5.36 p/unit.

(1 electrical unit = 1 kWh).

The relevant properties of the Freon 12 are as follows:

Press MN/m²	Sat. temp. °C	Spec. enthalpy kJ/kg h_f	h_g	Spec. entropy kJ/kgK s_f	s_g
0.362	5	40.7	189.7	0.1587	0.6942
0.745	30	64.6	199.6	0.2399	0.6854

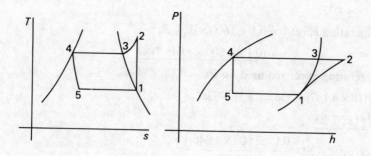

(a) $s_1 = \underline{0.6942 \text{ kJ/kgK}}$

$s_2 = s_1$

and $s_2 = s_3 + c_{pv} \ln \dfrac{T_2}{T_3}$

∴ $0.6942 = 0.6854 + 0.71 \ln \dfrac{T_2}{T_3}$

from which,

$$\ln\frac{T_2}{T_3} = \frac{0.6942 - 0.6854}{0.71} = \frac{0.0088}{0.71} = 0.0124$$

$$\therefore \frac{T_2}{T_3} = 1.0125 \quad \text{and} \quad T_3 = 30 + 273 = \underline{303 \text{ K}}$$

$$\therefore T_2 = 1.0125 \times 303 = \underline{306.8 \text{ K}}$$

$$h_2 = h_3 + c_{pv}(T_2 - T_3)$$
$$= 199.6 + 0.71(306.8 - 303)$$
$$= 199.6 + (0.71 \times 3.8) = 199.6 + 2.7$$
$$= \underline{202.3 \text{ kJ/kg}}$$

$$h_4 = \underline{64.6 \text{ kJ/kg}}$$

Heat transfer in condenser $= h_2 - h_4$
$$= 202.3 - 64.6$$
$$= \underline{137.7 \text{ kJ/kg}}$$

$$\therefore \text{mass flow of refrigerant} = \frac{50 \times 10^3}{137.7}$$
$$= \underline{363.1 \text{ kg/h}}$$

(b) $h_4 = h_5 = \underline{64.6 \text{ kJ/kg}}$

$$\therefore 64.6 = 40.7 + x_5(189.7 - 40.7)$$

from which,

$$x_5 = \frac{64.6 - 40.7}{189.7 - 40.7} = \frac{23.9}{149} = \underline{0.16} = \text{dryness fraction entering evaporator.}$$

(c) Specific work $= h_2 - h_1$
$$= 202.3 - 189.7 = \underline{12.6 \text{ kJ/kg}}$$

mass flow of refrigerant $= \dfrac{363.1}{3600} = \underline{0.1009 \text{ kg/s}}$

$$\therefore \text{power to driving motor} = \frac{12.6 \times 0.1009}{0.75}$$
$$= \underline{1.695 \text{ kW}}$$

(d) Cost of running/8 hour day
$$= 1.695 \times 8 \times 5.36$$
$$= \underline{72.7 \text{ p}}$$

Questions 6

1 A vapour compression refrigerator circulates 0.025 kg of Freon 12/s and operates between the pressure limits of 0.1508 MN/m² and 0.651 MN/m². At the commencement of compression the Freon 12 is dry saturated. There is 5 °C undercooling at the end of condensation. Take the specific heat capacity of superheated Freon 12 = 0.72 kJ/kgK and the specific heat capacity of liquid Freon 12 = 0.96 kJ/kgK. Determine:
(a) the coefficient of performance;
(b) the dryness fraction of the Freon 12 as it enters the evaporator;
(c) the work done/kg of Freon 12;
(d) the refrigerating effect/h;
(e) the effective swept volume of the compressor in m³/min.

The relevant properties of Freon 12 are as follows:

Press MN/m²	Sat. temp. °C	Spec. enthalpy kJ/kg		Spec. entropy kJ/kgK		Spec. vol. m³/kg	
		h_f	h_g	s_f	s_g	v_f	v_g
0.1508	−20	17.8	178.7	0.0731	0.7088	0.00069	0.1089
0.651	20	59.7	197.7	0.2239	0.6869	0.00076	0.0269

[(a) 5.87, (b) 0.23, (c) 25.4 kJ/kg, (d) 11.142 MJ/h, 0.163 m³/min.]

2 A heat pump uses ammonia between the pressure limits of 0.516 MN/m² and 1.554 MN/m². The mass flow of ammonia is 0.05 kg/s. The dryness fraction at entry to the compressor is 0.95. At the end of condensation the temperature of the ammonia is 32 °C. Take the specific heat capacity of liquid ammonia = 4.8 kJ/kgK and the specific heat capacity of superheated ammonia = 2.98 kJ/kgK. Determine:
(a) the heat transfer available from the condenser/h;
(b) the dryness fraction of the ammonia as it enters the evaporator;
(c) the power required to drive the heat pump if the overall efficiency of the compressor and electric driving motor is 70%;
(d) the cost of running the heat pump/12 hour day if the electrical cost is 5.5 p/unit.

The relevant properties of ammonia are as follows:

Press MN/m²	Sat. temp. °C	Spec. enthalpy kJ/kg		Spec. entropy kJ/kgK	
		h_f	h_g	s_f	s_g
0.516	5	204.5	1450	0.799	5.276
1.554	40	371.5	1473	1.360	4.877

[(a) 215.26 MJ/h, (b) 0.103, (c) 10.1 kW, (d) £6.67]

Index

Adiabatic
  process 10
    frictionless 27
    and reversibility 10
    and temperature-entropy chart 27
Air
  compressor 88
    rotary 43, 94
  engine 86
    reversed 89, 145
  standard cycle 68
  standard efficiency 68
Ambient temperature 141, 142

Bell–Coleman refrigerator 89, 145
Boiler 126
Brayton cycle 86
Brayton, George 86

Carnot cycle
  gas 68
  reversed 143, 147
  steam 128
Carnot efficiency 16, 72, 129
Carnot, Sadi 5
Carnot's principle 13
Clausius, Rudolf 5
Coefficient of performance 141
Composite cycle 115
Condensate 127
  extraction pump 127
Condenser 127
Constant pressure
  change of entropy 37
  cycle 86
  gas turbine 93, 94
Constant temperature
  change of entropy 37
  cycle 72

Constant volume
  change of entropy 37
  cycle 78
Cooling tower 127
Critical point 26, 33, 34
Cryogenics 141
Cycle 1
Cycles
  gas 65
  refrigeration 141
  steam 126

Diesel
  cycle 105
    high-speed 115
Diesel, Rudolph 105
Dual combustion cycle 115

Efficiency
  air standard 68
  isentropic 51
  ratio 66
  relative 66
  thermal 66
Energy
  non-flow equation 3
    and reversibility 11
  steady-flow equation 3, 51
Engine
  thermodynamic, principles of 13
Enthalpy
  -entropy chart, vapours 33
  -pressure chart, vapours 34
Entropy 17
  chart, gas 40
  -enthalpy chart, vapours 33
  gas 35
  liquid 20
  of evaporation 22
  superheated vapour 24

-temperature chart, vapours 25
vapours 20
Ericsson, John 86

Feed pump 127
First law of thermodynamics 1
  and the cycle 1
  and heat and work transfer 1
  and the non-flow energy equation 2
  and the steady-flow energy equation 2
Fuel, specific consumption 67

Gas
  adiabatic process 43
  cycles 65
  entropy chart 40
  entropy of 35
  turbine 55, 93

Heat pump 142, 151
High-speed Diesel cycle 115

Ideal steam plant cycles 126
Isentropic 27
  efficiency 51
Isothermal process
  and reversibility 10
  and the temperature-entropy chart 26

Joule
  air engine 88
  and internal energy 12
  cycle 86
Joule, James Prescott 86

Kelvin, Lord 6
Knocking 83

Law of thermodynamics
  first 1
  second 4

Mean effective pressure 67
Mollier chart 34

Non-flow energy equation 3
  and reversibility 11
Nozzle 53

Oil engine 105
Otto cycle 78
Otto, Dr N.A. 78

Pinking 83
Planck, Max 6
Polytropic process
  change of entropy 38
  and reversibility 11
Pressure-enthalpy chart, vapours 34

Rankine
  cycle 130
  efficiency 132, 133
Rankine, William John 132
Reference state for entropy 20, 40
Refrigerant 142
Refrigerating effect 141
Refrigeration 141
Refrigerator
  vapour compression 147
Refersed
  air engine 143, 147
  Carnot cycle 143, 147
Reversibility
  and the adiabatic process 10
  and the constant pressure process 11
  and the constant volume process 11
  and the isothermal process 10
  and the non-flow energy equation 11
  and the polytropic process 10
Rochas, Beau de 78
Rotary air compressor 93, 94

Second law of thermodynamics 4
  and heat and work transfer 4
  and temperature gradient 4
Steam
  entropy of 20
  plant cycles 126
  specific consumption 67
  turbine 52
Stirling cycle 123
Stirling, Dr Robert 123
Stirling, James 123
Superheat
  entropy of 24
Superheated steam 24

Temperature
  -entropy chart, vapours 25

Thermal efficiency 66
   actual 66
   ideal 66
   theoretical 66
Thermodynamic
   engine, principle of 13
   reversibility 8
Thermodynamics
   first law of 1
   second law of 4
Thompson, William 6
Throttling 148

Turbine
   gas 55, 93
   steam 126, 130, 131

Vapour
   cycles 65, 126
   entropy of 20

Work
   indicated 67
   ratio 67
   specific output 67